Limnological Reseach in India

– Editor –

S. R. Mishra

2018

Daya Publishing House®

A Division of

Astral International Pvt. Ltd.

New Delhi - 110 002

First Published, 1999
Reprinted, 2018

ISBN: 978-81-7035-204-4 (Hardbound)

Published by : **Daya Publishing House®**
A Division of
Astral International Pvt. Ltd.
– ISO 9001:2008 Certified Company –
4736/23, Ansari Road, Darya Ganj
New Delhi-110 002
Ph. 011-43549197, 23278134
E-mail: info@astralint.com
Website: www.astralint.com

Dedicated to my respected teacher
Prof. (Dr.) D.N. Saksena,
M.Sc., Ph.D., F.A.Z., F.A.E.B., Chairman,
Board of Studies, Jiwaji University, Gwalior,
Whose dedication and allegiance to education
is my inspiration.

– Editor

Preface

Limnology embraces the biochemical and geological characteristics of the fresh water organisms while fresh water ecology emphasizes the interrelationship between organisms and immediate environment. The subject matter of limnology derives knowledge from varied disciplines like physics, chemistry, biology, geology, geography etc. and involves a great deal of detailed field and laboratory studies to understand the structural and functional aspects of the fresh water environments from a holistic point of view.

The source and nature of fresh water, its motion and changing conditions as it flows to the sea and the life it support along the way, are now the subject of limnology, the science of fresh waters and the science of aquatic life. These sciences as they are known today are young disciplines, only about a century old, but they had their roots as did so much of modern science, with the philosophers of ancient Greece.

Limnology now includes study of running water as well as of standing water. This branch of science has attracted lot of attention during recent years and lot of high quality work is being done. The present volume is the outcome of a collaborative effort of the researchers and scientists in this discipline. It presents the subject on a wider perspective and provides upto-date information on various aspects of limnological research in India.

I am greatly indebted to all those who have contributed to this volume. My mother has always been a constant source of inspiration for me, I pay my sincere regards to her.

S.R. Mishra
Editor

Contents

Contributors

Alam, Md. Afaque	:	Fishery Science and Agriculture Unit, Department of Zoology, Aligarh Muslim University, Aligarh-202002 (U.P.)
Balasubramanian, S.	:	Department of Environmental Sciences Bharathiar University, Coimbatore-641046
Chowdhury, P. Basu	:	Department of Botany, Netaji Nagar, Day College, Regent Park, Calcutta-700040
Dave, R.K.	:	Department of Zoology, Govt. P.G. College, Dhar (M.P.)
Dhakad, N.K.	:	Department of Zoology, Govt. P.G. Girls College, Indore (M.P.)
Duda, P.L.	:	Department of Biosciences, University of Jammu (J&K)-180004, India
Gaur, Rajeev K.	:	Fishery Science and Agriculture Unit, Department of Zoology, Aligarh Muslim University, Aligarh-202002 (U.P.)
Gautam, Ashutosh	:	Indian Glycols Ltd., A-1, Industrial Area, Bazpur Road, P.O. Box-22, Kashipur-244713 (U.P.)
Geetha, P.	:	Department of Environmental Sciences, Bharathiar University, Coimbatore-641046
Jaganathan, R.	:	Department of Zoology, Kongunadu Arts and Science College, Coimbatore-641029
Khan, Asif A.	:	Fishery Science and Agriculture Unit, Department of Zoology, Aligarh Muslim University, Aligarh-202002 (U.P.)

Maria Charles, P.	:	Department of Zoology, Govt. Arts College (Men,) Nandanam, Madras-600035
Mishra, Awadhesh Kumar	:	Zoology Department, B.M.D. College, Dayalpur, BRA Bihar University, Vaishali-844502 (Bihar)
Mishra, S.R.	:	30, Geeta Bhawan Road, Bhind (M.P.) - 477001
More, Y.S.	:	Department of Botany, K.A.M.P. and Kai N.K.P. Senior College, Pimpalner, Distt. Dhule-424306 (MS)
Murugesan, A.G.	:	Manonmaniam Sundaranar University, Sri Paramakalyani Centre for Environ. Sci., Alwarkurichi-627412 (TN)
Nair, Anita	:	Department of Applied Science (Pulp & Paper), B.N. Degree College, Dandeli-581325 (N.K.)
Nandan, S.N.	:	P.G. Department of Botany, S.S.V.P.S.'s L.K. Dr. P.R. Ghogrey Science College, Dhule-425005 (MS)
Padmalatha, G.	:	Department of Environmental Sciences, Bharathiar University, Coimbatore
Pandit, Ashok K.	:	Centre of Research for Development, University of Kashmir, Srinagar-190006 (India)
Prakash, M.M.	:	Department of Zoology, Govt. P.G. College, Jhabua (M.P.)
Rajput, Sushma	:	Bacteriology Laboratory, Department of Biosciences, R.D. University, Jabalpur-482001 (M.P.)
Ramanibai, R.	:	Department of Zoology, University of Madras, Guindy Campus, Madras-600025
Sahi, D.N.	:	Department of Biosciences, University of Jammu (J&K)-180004, India
Sai Sastry, A.G.R.	:	Department of Zoology, A.N.R. College, Gudivada-521301 (A.P.)

Sarkar, S.K. : Department of Zoology, Netaji Nagar Day College, Regent Park, Calcutta-700040

Sasikala, D. : Department of Zoology, University of Madras, Guindy Campus, Madras-600025

Shanthi, K. : Department of Zoology, Kongunadu Arts & Science College, Coimbatore-641029

Sharma, Anjana : Bacteriology Laboratory, Department of Biosciences, R.D. University, Jabalpur-482001 (M.P.)

Shenvi, A.K. : Department of Applied Science (Pulp & Paper), B.N. Degree College, Dandeli-581325 (N.K.)

Sivasubramani, R. : H.K.R.H. College, Uthamapatayam-625533 Theni district, Tamil Nadu

Sukumaran, N. : Manonmaniam Sundaranar University, Sri Pramakalyani Centre for Environ. Sci., Alwarkurichi-627412 (TN)

Suresh, B. : J.S.S. College of Pharmacy, Ooty.

Venkatesan, T. : Department of Environmental Sciences, Bharathiar University, Coimbatore-641046

Verma, A.K. : Rural Technology Unit, Regional Research Laboratory (C.S.I.R.), Canal Road, Jammu Tawi-180001, India.

Vipradas, K.C. : Taraporevala – Marine Biological Research Station, N.S. Road, Mumbai-400002

1
Water Quality of River Periyar (River Suruliyar) in Tamil Nadu

R. Sivasubramani

INTRODUCTION

There are only a few reports on the assessment of water quality of Southern Indian rivers. These include the assessment of water quality of the Adayar river by Chacko and Ganapathi (1945), River Vaigai by Anandhavalli Mahadevan and Krishnaswamy (1980), Rivers Periyar in Kerala by Joy *et al.*, (1990), Rivers Godavari, Krishna and Tungabhadra by Ramesh *et al.*, (1992). River Periyar (approximately in north latitude 9°10' and at altitude of 5000 ft.) is the largest among the rivers flowing west through Kerala Sate and hence the name Periyar river. The rainfall in this region ranges from 3000 mm to 4000 mm which augments the flow of water into the River Periyar. The water drawn from Periyar lake through the approach channel and tunnel is let into the tributary, Vairavanar. The flow then runs through Suruliyar in the Cumbum valley in Tamil Nadu. This has become the route for the Periyar water in Tamil Nadu. The River Periyar in Tamil Nadu is the major water source of the Cumbum valley. The River water is widely used for bathing, washing and religious purposes. It receives effluents, and runoff from the villages and towns along its banks.

In the present study, various parameters like physico-chemical parameters, microbiological parameters, physical features and hydrological observations were chosen to assess the quality of water.

MATERIALS AND METHODS

Study Area

The study area was divided into 9 sampling stations (Fig. 1). Water samples were collected from nine sampling stations from Periyar lake to Theni Aranmanai Pudur (where River Periyar joins Theni river) within the stretch of 53.5 km. There were three upstream and six downstream stations among these nine sampling stations. The three upstream station water samples were analysed for their background quality.

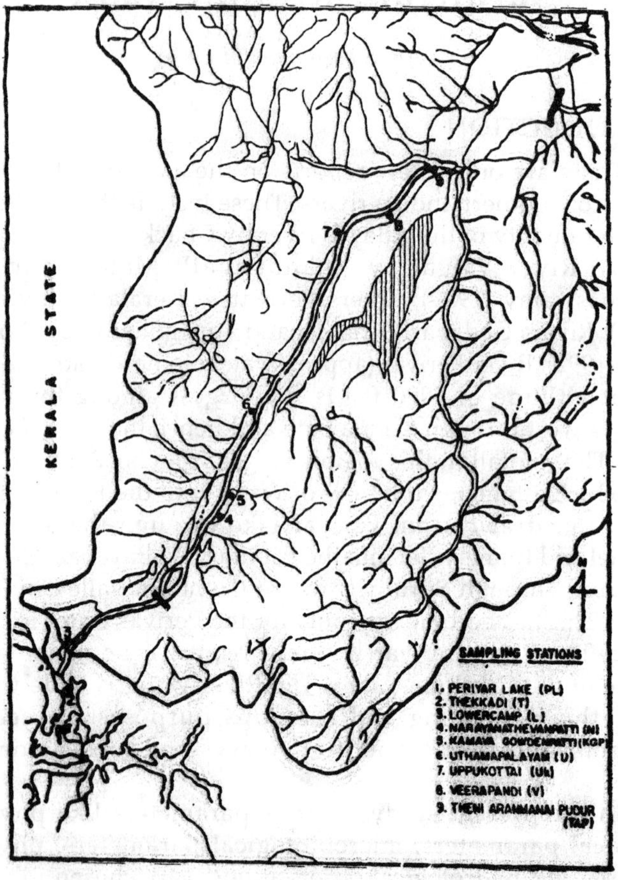

FIG. 1 : MAP OF RIVER PERIYAR (RIVER SURULIYAR) IN TAMIL NADU

Hydrology

For hydrological observations, the flow measurement data and rainfall data were collected from Public Works Department (PWD), Uthamapalayam. A hydrograph was constructed from the monthly mean flow of the river, to know the fluctuation of the flow during the study period.

Physico-Chemical Analyses

For physico-chemical analyses, thirteen physico-chemical parameters were estimated in accordance with the standard methods found in APHA (1985) and Trivedy and Goel (1984). The collected samples were carefully carried in pre-cleaned polythene containers. Using portable pH meter and Winkler's method an on-the-spot study was carried out to determine pH and dissolved Oxygen concentration respectively. For a period of two years samples from all the stations were collected and analysed every week.

Microbiological Analyses

For bacterial analyses, water samples were taken in pre-sterilised and paper wrapped BOD bottles in ice-boxes to the laboratory. In bacterial analyses most probable number (MPN) of coliforms and MPN of faecal coliforms were estimated by multiple tube fermentaton techniques according to the Standard methods (APHA, 1985). Most probable number of coliforms was calculated from observed combination of positive tubes. The values were calculated using MPN tables.

RESULTS AND DISCUSSION

Table 3 provides the list of various sources of pollution of the sampling stations. The PL, T, and KGP are the stations exposed to human impact to a very negligible extent. That stations L, N, UK, and V are exposed to intermittent human impact, whereas stations U and TAP are exposed to a very high frequency of human impact.

Hydrological findings

Hydrograph – Fig. 2 shows the fluctuations of flow during 1989-91 in the River Periyar in Tamil Nadu.

The daily flow range varied from 0.02 m³/sec. to a high of 5.4 m³/sec. Owing to high velocity winds, during the months of August and November high peaks of flow were recorded. Sometimes the flow rate was below the average.

Fig. 3 shows rainfall during 1989-91 varies from nil to 31.5 mm. Rainy season includes South-West and North-East monsoons. South-West monsoon starts from June and ends in September and North-East monsoon from October to December. Maximum rainfall is observed in June and July in the upstream stations (PL and T) from South-West monsoon. Rainfall and river discharge are closely related.

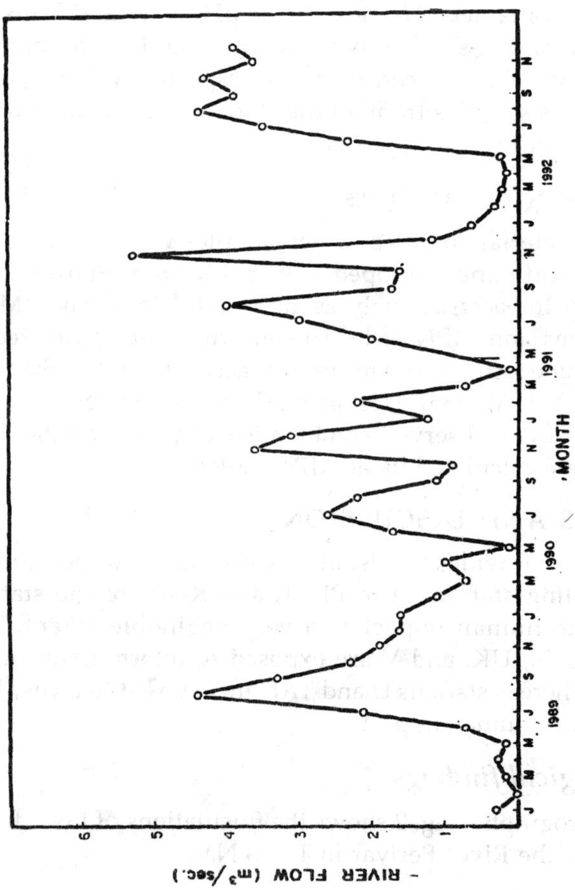

FIG. 2 : HYDROGRAPH OF RIVER PERIYAR (RIVER SURULIYAR) IN TAMIL NADU

Physico-Chemical quality

Water was bright and clear at stations PL, T, N, KGP, UK and V. There were no fluffy deposits on stones, in the first three stations. In the remaining three stations, very light deposits on

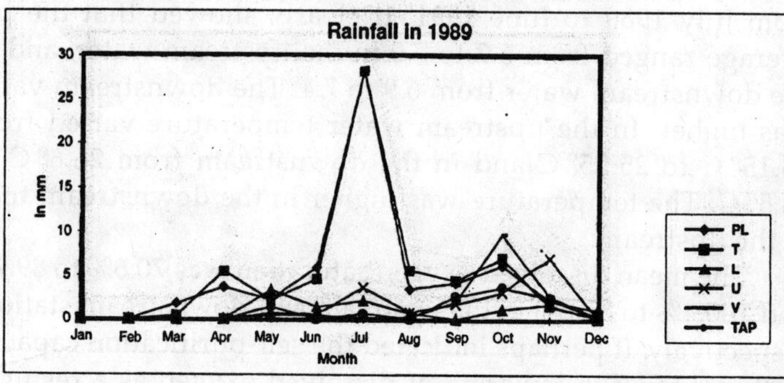

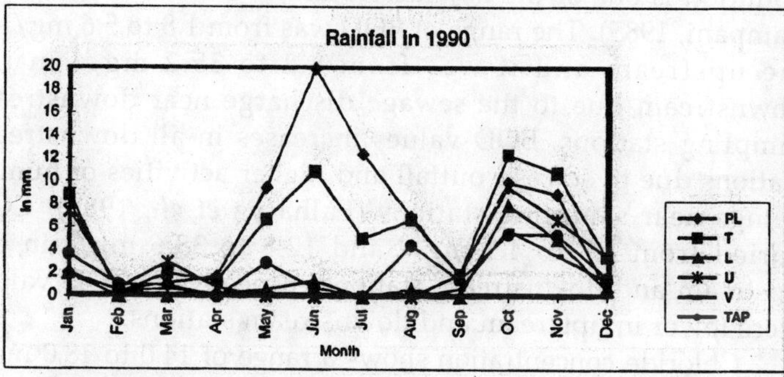

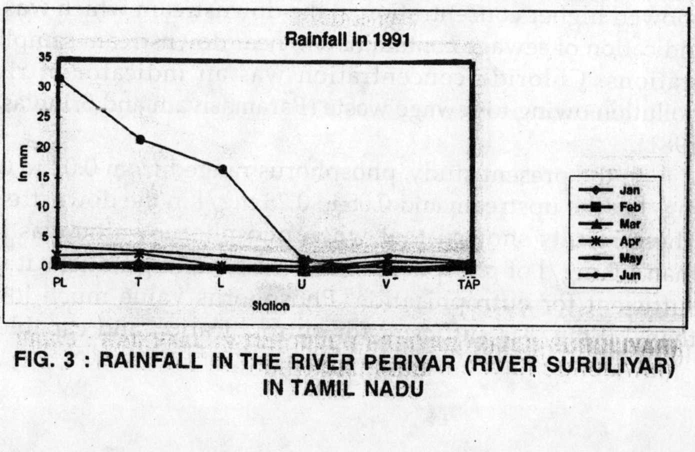

FIG. 3 : RAINFALL IN THE RIVER PERIYAR (RIVER SURULIYAR) IN TAMIL NADU

stones were noticed. Faecal smell emanated from the water at station TAP. It was also turbid. There were light coatings of sewage fungus on the stones.

Table 1 provides the mean and standard deviation of various physico-chemical parameters of all the upstream and downstream sampling stations of the River Periyar for two years from July 1989 to June 1991. It clearly showed that the pH average ranged from 6.7 to 7.0 in the upstream water, and in the downstream water from 6.9 to 7.4. The downstream value was higher. In the upstream water temperature varied from 23.15° C to 25.25° C and in the downstream from 26.6° C to 28.5° C. The temperature was higher in the downstream than in the upstream.

The mean dissolved oxygen saturation was 70.6% to 89.7% and 100.4% to 82.6% in the upstream and downstream stations respectively. It perhaps indicated the self-purification capacity of this river. The lowering of dissolved oxygen as a result of pollution is one of the key factors in the distribution of biota (Jampani, 1985). The range of BOD was from 1.8 to 5.6 mg/l in the upstream and it was from 8.2 to 35.2 mg/l in the downstream due to the sewage discharge near downstream sampling stations. BOD values increases in all downstream stations due to sewage outfall and higher activities of human beings near sampling stations (Palharya *et al.*, 1988). COD varied from 1.7 to 4.1 mg/l and 12.8 to 38.6 mg/l in the upstream and downstream stations respectively. COD values were lower in upstream and downstream stations.

Chloride concentration shows a range of 14.0 to 18.0 mg/l in the upstream and 18.7 to 48.0 mg/l in the downstream. It showed higher concentration in the downstream which was an indication of sewage contamination near downstream sampling stations. Chloride concentration was an indicator of river pollution owing to sewage waste (Paramasivam and Srinivasan, 1981).

In the present study, phosphorus ranged from 0.02 to 0.04 mg/l in the upstream and 0.05 to 0.23 mg/l in the downstream. These results showed that when phosphorus value was less than 1.0 mg/l of phosphorus in all the sampling stations it was sufficient for eutrophication. Phosphorus value much lower than 1.0 mg/l is sufficient for eutrophication and capable of

TABLE 1 : PHYSICO-CHEMICAL PARAMETERS OF RIVER PERIYAR (RIVER SURULIYAR) IN TAMIL NADU AT DIFFERENT SAMPLING STATIONS

Parameters	Upstream Stations				Downstream Stations					Permissible limit	
	PL	T	L	N	KGP	U	UK	V	TAP	Drinking Water (ISI 1982)	Bath Water (BIS, 1982)
pH	6.7±0.1	6.9±0.2	7.0±0.2	6.9±0.1	7.1±0.2	7.3±0.3	7.1±0.3	7.2±0.2	7.4±0.3	6.5-8.5	6.5-8.5
Air Temp. (°C)	24.34±3.6	24.97±3.6	28.16±3.1	28.9±2.3	29.6±2.5	29.8±2.5	30.5±2.1	31.18±2.0	31.80±2.9	–	–
Water Temp.(°C)	23.15±3.5	23.35±3.4	25.25±3.0	26.62±2.8	25.36±2.8	25.87±3.0	25.44±3.0	27.74±2.3	28.47±2.9	40°C	–
DO% saturation	70.6±21.7	71.7±22.3	89.7±30.3	100.4±19.0	92.9±23.0	84.3±25.0	86.3±29.3	86.9±26.2	82.6±35.3	6	5
BOD	1.8±0.7	2.0±0.5	5.6±2.9	8.2±4.5	9.6±4.7	12.3±8.2	8.7±4.9	9.3±5.1	35.2±21.7	2	3
COD	1.7±0.9	1.8±1.0	4.1±1.6	12.8±3.9	14.2±5.0	15.7±6.0	11.1±4.1	14.2±5.2	38.6±22.2	250	–
Nitrites	0.01±0.00	0.00±0.00	0.02±0.01	0.13±0.08	0.10±0.06	0.20±0.12	0.12±0.07	0.14±0.09	0.37±0.19	–	–
Chlorides	14.0±2.4	15.0±2.9	18.0±3.0	18.7±3.1	20.1±7.2	28.5±7.0	19.0±4.1	22.7±5.9	48.0±16.7	250	–
Phosphates	0.02±0.01	0.03±0.01	0.04±0.02	0.05±0.02	0.04±0.03	0.14±0.09	0.05±0.03	0.06±0.05	0.23±0.20	–	–
Conductivity (μmhos/cm)	80.0±12.5	100.8±14.1	110.0±14.9	200.0±45.2	180.6±26.2	230.2±33.0	170.5±24.5	210.7±26.7	350.9±151.1	–	–
Total solids	13.2±1.2	17.2±1.5	20.8±2.3	150.3±17.2	100.2±20.2	225.1±60.1	101.8±19.5	165.6±62.7	260.8±108.0	500	–
Total Alkalinity	27.5±12.9	29.3±11.1	33.2±13.2	40.0±23.1	48.6±25.5	51.2±25.8	46.2±23.8	40.1±26.2	60.7±26.0	120	–
Total hardness	7.1±2.0	9.2±2.7	18.2±4.3	90.5±13.2	120.1±15.3	177.3±20.3	130.7±18.5	145.2±23.2	220.1±58.8	100	–

All the values are in mg/l except pH, Temperature, Dissolved Oxygen and Conductivity. Each value represents the mean of 57 samples.

supporting a rich plankton flora (Vollenweidder, 1968). Nitrites varied from 0.01 to 0.02 mg/l in the upstream and 0.13 to 0.37 mg/l in the downstream and the value was higher in the downstream stations.

Conductivity shows a range of 80.0 to 110.0 µmhos in the upstream and 200.0 to 350.9 µmhos in the downstream stations. It was higher in the downstream stations. Total solids ranged from 13.2 to 28.8 mg/l in the upstream and from 150.3 to 260.8 mg/l in the downstream stations. The value was higher in the downstream stations than in the upstream stations.

Toal hardness varied from 7.1 to 18.2 mg/l in the upstream and 90.5 to 220.1 mg/l in the downstream. The high total hardness in some of the downstream stations was due to high sewage contamination. The hardness of river water was of considerable significance in connection with the discharge of sewage effluents (Rai, 1974). Total alkalinity ranged from 25.7 to 33.2 mg/l in the upstream and 40.0 to 60.7 mg/l in the downstream. Alkalinity was higher in the downstream than in the upstream.

Among the downstream stations, difference in the level of physico-chemical parameters between stations depended upon the sources of pollution (Table 3).

Microbiological Quality

Bacteriological parameters were of great importance in the hygienic evaluation of water (Bonde, 1977). Bacterial contamination in different water bodies had been reported by different authors (Bhattacharjee *et al.*, 1989; Panda *et al.*, 1993). Table 2 provides the means of MPN of total coliforms and MPN of faecal coliforms of nine sampling stations under the present study. It showed that the coliforms count was minimum (720.9/100 ml) in the station PL and the maximum (12,153/100 ml) in the station TAP. The increases in the MPN of coliforms and faecal coliforms at the sampling U and TAP were more marked than at the other upstream and downstream stations. From sampling station N to TAP downstream there was a steady rise in the MPN of total coliform bacterial quality.

In bacterial populations, the faecal coliform count was not always so high as the total coliform count (Abhaykumar and Dube, 1992). It was found that in all the sampling stations in

TABLE 2 : BACTERIAL QUALITY OF RIVER PERIYAR (RIVER SURULIYAR) IN TAMIL NADU AT DIFFERENT SAMPLING STATIONS

Parameters	Upstream Stations					Downstream Stations				Permissible limit		
										Drinking Water		Bath Water
	PL	T	L	N	KGP	U	UK	V	TAP	WHO (1984)	ISI (1982)	BIS (1982)
MPN of Coliform (Geometric mean no./ 100 ml.)	720.0	1235.7	1626.9	1505.7	2123.5	10553.7	2873.8	3623.9	12153.1	0/100ml	1/100ml	500
MPN of faecal coliform (Geometric mean no./ 100 ml.)	9.4	53.5	307.3	439.3	627.6	3726.1	733.1	973.1	4920.1	–	–	100

Each Value represents the mean of 57 samples.

TABLE 3 : SOURCES OF POLLUTION OF THE SAMPLING STATIONS

Soures of Pollution	Stations (Downstream)					
	N	KGP	U	UK	V	TAP
Open drainage opening into the water	—	—	000	—	—	000
Defecation on the banks	0	0	00	0	00	000
Washing clothes	—	0	00	0	0	000
Bathing	0	0	00	0	00	000
Cattle bathing	—	0	0	0	0	00
Animal waste	—	—	0	—	—	000
Solid waste dumping	—	—	0	—	0	000
Land runoff	0	0	00	0	0	00
Urban runoff	—	—	0	—	0	000

— **Nil**
0 **Negligible**
00 **Intermittent**
000 **High**

the River Periyar in Tamil Nadu, total coliform count was higher than that of faecal coliform. There was progressive increase in the faecal coliform in the downstream station N to TAP (except at N and UK). This indicated a strong faecal pollution of the entire downstream stretch due to bathing and defecation on the banks. In the higher bather load days there was temporal relationship between bather load and water quality indicator levels (faecal pollution) (James, 1986).

Criteria for raw water supply

The Periyar is the source of water supply for Cumbum valley. Water for drinking purpose is collected through infiltration galleries from station L to TAP. Some villagers living adjacent to the sampling stations fetch drinking water from the river itself.

Among the physico-chemical parameters analysed, only pH, dissolved oxygen, chlorides, total solids, and total alkalinity confirm to the standard limit (ISI, 1982). At all the stations the values of organic pollution indicating parameters such as BOD exceeded the permissible limit but COD values were below the permissible limit.

As per the World Health Organisation standard (WHO, 1984) of raw water supply, water should contain less than 0/100 ml of MPN of coliform; European Economic Community (EEC, 1980) has prescribed 0/100 ml of MPN of coliform whereas 1/100 ml of MPN of coliform has been prescribed as the standard by the Indian Council of Medical Research (ICMR). The MPN of coliform per 100 ml in the samples collected at all the stations exceeded the prescribed limit of the above said standards.

Hence in all the sampling stations the river water was found heavily contaminated. So the water could be used for drinking purposes only after applying conventional methods of pre-treatment.

Criteria for bathing water

The levels of parameters for bathing water were shown in Table 1 & 2. When the coliform count is 500/100 ml. or less, water is deemed suitable for bathing (BIS, 1992). The coliform count in all the upstream and downstream stations in the Periyar river was above 500 coliform/100 ml, which exceeded the permissible limit. So the water was not suitable for bathing purpose.

Criteria for oxygen demanding wastes

BOD is an essential parameter in stream pollution control management and in the evaluation of the self purification capacity of streams (Trivedy and Goel, 1984). The Royal Commission on sewage disposal (1912) has suggested a 5 day, BOD test for rivers and streams. The permissible limit of BOD for raw surface water is 2 mg/l (ISI, 1982) and it is prescribed as 5 mg/l by United States Public Health Service (USPHS).

Water samples from all the stations exceeded the permissible limit of the above said standards for oxygen demanding wastes.

There should be an adequate quantum of dissolved oxygen in a stream (White and Dracup, 1977). According to the Indian

Standard (ISI, 1982) the permissible limit for raw water supply is 6 mg/l of dissolved oxygen while the prescription of the European Economic Community standard (EEC, 1980) is 5 mg/l and the Indian Council of Medical Research standard is 5 mg/l of oxygen. The dissolved oxygen saturation was above 70 per cent in stations PL and T and it was above 80 percent in station L and in the downstream stations. So, the level of DO saturation in all the stations conform to the Indian Standard, the European Economic Community Standard and the Indian Council of Medical Research Standard.

The river with self-purification capacity could assimilate the existing load (Lester, 1967). The self-purification capacity of the river was borneout by the average presence of 84.3 percent and 82.6 percent of dissolved oxygen saturation even at the highly polluted stations U and TAP.

Though the Periyar river water satisfied parameters like pH, total solids, chlorides, total alkalinity, DO, it did not satisfy the other water quality parameters. Hence, disinfection was the only pre-treatment (by chlorination) to be undertaken before water was supplied for human consumption.

CONCLUSION

Owing to urbanization, the quality of Periyar river water has deteriorated. Pollution parameters varied from station to station because of the quality and quantity of domestic wastes around the sampling stations. It is the bounden duty of the civic authorities to check the direct mixing of the sewage water and domestic wastes with the waters of this river. Eventhough it receives no industrial effluents it could be ranked as one of the moderately polluted rivers in the country.

SUMMARY

The present study deals with the Water Quality of River Periyar (River Suruliyar) in Tamil Nadu. The monitoring of water quality of River Periyar (River Suruliyar) in Tamil Nadu was carried out for a period of two years i.e. from July 1989 to June 1991. The important physico-chemical and bacterial parameters were analysed. To assess the quality of water at various sampling stations, along the River Periyar the levels of various observed parameters have been compared with the relevant standards. The quality of the Periyar river water was also undertaken for discussion. In many sampling stations, the quality of the river water examined was found not within the permissible limits for human consumption.

ACKNOWLEDGEMENTS

The author is thankful to the Management and Principal of H.K.R.H. College, Uthampalayam and U.G.C., New Delhi, for encouragement and facilities provided to conduct this work.

REFERENCES

Abhaykumar, V.K. and Dube, H.C. 1992. Occurrence and distribution of bacterial indicators of faecal pollution in the tidal waters of a muddy coast. Poll. Res. **11**(4): 235-238.

APHA 1985. Standard Methods for the examination of water and waste water, APHA, AWWA, WPCF, 16th Ed. Washington D.C.

Anandhavalli, A. 1980. Biological and physico-chemical indicators of polluted waters, in and around Madurai. Ph.D. thesis, Madurai Kamaraj University.

BIS 1982. Standard tolerance limits for bathing waters. Bureau of Indian Standards. IS: 2296.

Bhattacherjee, J.W., Rameteke, P.W., Pathak, S.P. and Ray, P.K. 1989. Bacteriological quality of drinking water in rural India. J. Indian Wat. Works, Association, April-June, 185-190.

Bonde, G.J. 1977. Bacterial indication of water pollution. In Droop, M.R. and Januasch, H.W. Advances in aquatic microbiology, 273-364. London and New York, Academic Press.

Chacko, P.I. and Ganapathi, S.V. 1945. Some observations on the Adayar river with special reference to its hydrobiological conditions. J. Indian Geogra. **24**(3): 35-49.

EEC 1980. Council directive of 15 July 1980. Official Journal of the European Economic Community, **30**: 8-80 (August 1980).

ISI 1982. Tolerance Limits for Inland water subject of Pollution, Indian Standard Institutions, New Delhi.

James, P.S. 1986. Temporal distribution of faecal pollution indicators and opportunistic pathogens at a lake Ontario (Canada), bathing beach, J. Great Lakes Res. 12(3): 154-160.

Jampani, CSR, Kumari, D.S., Sudha, K. and Nalini, P. 1985. Effect of Diary effluent of algae. Proc. Nat. Symp. Pure and Appl. Limno (ed) Adoni, A.D. Sull. Bot. Soc. Sagar. **32**: 183-186.

Joy, C.M., Balakrishnan, K.P. and Joseph, A. 1990. Effect of Industrial discharges in the ecology of phytoplankton production in the River Periyar (India). Water Res. **24**(6): 787-796.

Lester, W.F. 1967. 'Management of river water quality', In: River Management (Ed. Isaac, P.C.G.), Maclaren and Sons Ltd., London.

Palharya, J.P. and Malviya, S. 1988. Pollution of the Narmada at Hoshangabad in M.P. and suggested measures for control. pp. 55-86. In: (R.K. Trivedy Ed). Ecology and Pollution of Indian Rivers. Ashiah Publishing House, New Delhi.

Panda, S.K. and Dash, M.C. 1993. Water quality of some community ponds of Bargath town. Poll. Res. 12(1): 35-47.

Paramsivam, M. and Sreenivasan, A. 1981. Changes in algal flora due to pollution in Cauvery river. Indian J. Environ. Hlth. **23**(3): 222-238.

Rai, H. 1974. Limnological Studies on the River Yamuna at Delhi, India, Part - I Relation between the chemistry and state of pollution in the River Yamuna. Arch. Hydrobiol. **73**: 369-393.

Ramesh, N.B. Srikanth, R. and Rao, A.M. 1992. Evaluation of water quality in three rivers of Andhra Pradesh. Poll. Res. **11**(4): 209-212.

Royal Commission on Sewage Disposal 1912. Eighth report. H.M. Stationery office, London.

Trivedy, R.K. and Goel, P.K. 1984. Chemical and Biological methods for water pollution studies. Environ. Publications, Karad, Maharashtra (India).

Vollenweidder, R.A. 1968. Scientific fundamentals of the Eutrophication of lakes and flowing waters with special reference to nitrogen and phosphorus as factors of eutrophication. O.E.C.D., Paris, Report No. DAS/CSI/18-27.

White, J.D. and Dracup, J.A. 1977. 'Water quality modelling of a high mountain'. JWPCF, 49, 2179.

WHO, 1984. Guidelines for drinking water quality. Vol. 1 and 2, Geneva, Switzerland.

2
Limnological Studies of Indian Rivers with Relation to Algae – A Review

S.N. Nandan & Y.S. More

INTRODUCTION

The limnological study is mainly concerning with the study of micro and macro-organisms of biological interest. But all the limnological groups can not be studied by single worker. In India, the various aquatic habitats with relation to algae were not studied extensively. For the study of water quality, the limnological observations are needed. As the living organisms found in water are reliable indicators of water quality. Most of the organisms are being extensively used as indicators of water pollution. Now a days, the aquatic organisms are called as "biological litmus papers." The biological indicators provide a direct clue and quick informations of the aquatic systems.

In U.S.A. and European countries limnological studies of rivers with relation to algae have been carried out extensively by earlier workers (Kofoid, 1903; Fritsch, 1903; Allen, 1920; Classen, 1927; Reinhard, 1931; Rice, 1938; Butchner, 1947; Fijerdingstad, 1950; Lacky and Hupp, 1956; Blum, 1956; Klein, 1957; Palmer, 1959; Waslik, 1965; Weber and Moore, 1967; Lam, 1971; Lack, 1971; Willem *et. al.*, 1942; Rzoska, 1976; Vanladingham, 1976; Aykulu, 1978; Adebisi, 1981; Holmes and Whitton, 1981; Yoshio and Ichinura, 1984; Brooker and Johnson, 1984; De Smet *et. al.*, 1987; Irmer, 1992; Hussain and Saudi, 1996).

REVIEW : ALGAE OF INDIAN RIVERS

The number of publications on limnology and ecology of fresh water bodies have been appeared in India. Though many rivers of India were studied hydrobiologically very few were studied with relation to algae.

Ghousudhin (1934) studied on algal flora of river Moosi. Biswas (1942) studied algal communities of river Hooghly to Calcutta. Bhimachar and David (1946) observed the effect of factory effluents on the Bhadra river fisheries at Bhadravati, Ganapati and Alikunhi (1950) observed effluents from the Mettur Chemical and Industrial Corporation Ltd. Mettur Dam of Madras and their pollutional effects on the fisheries of river Cauvery. Iyengar and Venkataraman (1951) observed seasonal succession of the Cooum river at Madras with special references to Diatometaceae. The rivers of upper plains which is branch of western ghats shooting into the Madurai district of Tamil Nadu were studied viz. (i) Pambor river, (ii) Parappor river, (iii) Kodianthanar river, (iv) Gundar river.

The physical, chemical and biological conditions of these rivers were studied by Ganpati and Chacko (1951a). Ganpati and Chacko (1951b) showed the effects of pollution on River Godavari due to the wastes of paper mills at Rajahmundry. Chacko and Ganpati (1952) studied hydrobiological survey of the Surali river. Chacko *et. al.* (1953) studied hydrobiology of the Malampuzha river. The hydrobiological studies on major rivers like the Godavari, Tungabhadra, Krishna and Cauvery were carried out in Madras State at present it is Tamilnadu State. Roy (1955) studied plankton ecology of River Hooghly at Patana, West Bengal.

Das and Shrivastava (1956, 59) observed seasonal fluctuations of phytoplankton composition of fresh water bodies. Motwani *et. al* (1956) made some important observations on the pollution of River Sone by factory effluents of the Rohtas industries at Dalmianagar, Bihar. Bhaskaran (1959) investigated the effects of Industrial wastes on river pollution in Bihar and Uttar Pradesh. Chakrabarty *et. al.* (1959) studied plankton and physico-chemical conditions on River Jamuna at Allahabad. Quasim and Siddiqui (1960) made preliminary observations of River Kali which is covered by the effluents of industrial waters. Rai (1962, 1974a, b) studied the hydrobiological aspects of river Yamuna at Okhla Delhi. David (1963) studied pollution of Bhadra, Godavari and Krishna rivers. He made the observations of dynamics of potamo plankton populations and also showed the relationship between the chemistry and the state of pollution in the river Yamuna. One of the largest rivers in India, the Ganga was studied by Laxminarayan (1965a, b) at Varanasi and Dehradun. Balusu *et. al.* (1967) have made certain observations on self purification of Khan river and its effect on Kshipra river. In Kerala, Verkey and Alexander (1968) have made preliminary report on the hydrobiology of Beypore river for the year 1964. Venkateswarlu (1969a, b, c 1986) carried out an ecological study of the algae of the Moosi river, Hyderabad with special reference to water pollution. In his investigation he determined the physico-chemical complexes, factory influencing the distribution of algae and periodicity. Rai (1978) made observations on algal communities of the Ganges river at Varanasi. Bapat and Madalpure (1971) carried out the work on hydrology of the river Kham in Maharashtra. Varma and Dalela (1975) studied on pollution of Kalinadi by industrial waste and showed biological index of pollution.

The Chambal river and its tributaries were studied at Kota in Rajasthan (Olaniya *et. al* 1976). Rama Rao *et. al.* (1978) observed the effects of pollution on River Khan (Indore) and made assessment of river water by biological community. Vass *et. al* (1977) carried out hydrobiological studies of algae from Zhelum river.

Govindan and Sundaresan (1979) observed seasonal succession of algal flora in polluted region of the Adyar river. Phytoplankton constituents were used as indicators of water quality in the study of the Adyar river (Jeeji Bai and Rajendran 1980). Zingde *et. al.* (1980) studied effects of industrial water disposal on the water quality of the river Kolak (Gujarat). Prasad and Singh (1980) have made contribution on biological markers of water pollution of Gomati river. Sreenivasan *et. al* (1980) made observations on pollution of River Cauvery due to industrial and urban wastes. Prasad and Saxena (1980) studied the blue-green algae in relation to industrial pollution of the river Gomati, Lucknow. They used the species of blue-greens as Marker species or 'Index species'. Gunale and Balakrishnan (1981) have used the algae as Biomonitors of eutrophication in the Pavana, Mula and Mutha rivers flowing through Poona city (Maharashtra). Rishi and Kachrao (1981) studied the effect of sewage and fertilizers on phytoplankton of the Doodhganga river (Kashmir). Barahate and Tarar (1981) studied algal flora of Tapti river, Bhusawal.

Pingale (1981) studied on ecology of algae of Pam river in Poona (Maharashtra) and assessed the organic pollution of river by using Palmer's Index of pollution. Paramsivam and Sreenivasan (1981) observed the changes in algal flora due to pollution in Cauvery river. Prasad and Singh (1982) used diatoms as indicators of water pollution of Gomati river, Lucknow. Patra and Nayak (1982) investigated limnobiotic survey of Mahanadi during winter season only. Boralkar *et. al* (1982) studied on soil pollution among the Krishna river in the state of Maharashtra.

Venkateswarlu and Kumar (1982) studied chemical and biological assessment of pollution in the river Moosi, Hyderabad (A.P.) Laal *et. al.* (1982) showed species diversity of periphyton as an index of pollution of Ganga river. Nandan and Patel (1983, 84, 85, 88, 90, 92) observed algal communities of river Vishwamitri flowing through Baroda city, as indicators of organic pollution. Agrawal (1983) studied Chambal river at Kota. Jagdale *et. al.* (1984) studied pollution of Godavari river water at Nanded. Chattopadhya *et. al.* (1984) studied pollutional status of river Ganga in Kanpur region. Raina *et. al.* (1984) showed pollution studies on river Jhelum. Sankaran (1984) studied the algae of Aliyor river. Somasheker (1984) showed phytoplankton constituents as indicator of water quality of Cauvery river and also made observations of algal flora of Kapila river. Venkateswarlu and Reddy (1985, 1986) observed algae as biomonitors in river ecology and also studied ecology of algae in paper mill effluents and their impact on the river Tungabhadra. Kar *et. al.* (1987) studied pollution studied in river IbIII: plankton pollution and primary productivity. Dora and Roy (1987) have made investigation of water quality of Subernarekha river. Sengupta *et. al.* (1988) studied inorganic pollutants of Ganga water in the Region of Berhampore to Katwa, West Bengal. Nautiyal (1988) studied Maneri Dam of river Bhagirathi–A Lacustrine environment. Mahajan (1988) has made observations of deterioration of Indian rivers.

Palharya and Malviya (1988) studied pollution of Narmada river at Hoshangabad in M.P. Shah (1988) observed physico-chemical aspects of pollution in river Jhelum (Kashmir) and Shamsunder (1988) studied the same river to assess water quality. Later on the same river was studied by Tiwari (1989). Sinha (1988) observed the effect of waste disposal on water quality of river Damodar Bihar. Shukla *et. al.* (1989) studied physico-chemical and Biological characteristics of river Ganga from Mirzapur to Ballia.

Sudhakar and Venkateswarlu (1989) showed the observations on ecological imbalances in the rivers of Tungabhadra and Godavari of Andra Pradesh. Chopra *et. al.* (1990) studied abiotic variability and primary productivity of River Yamuna at Naugaon, Uttarkashi-Garhwal. Preliminary studies on the water pollution of river Nandira was made by Tripathy and Adhikary (1990).

Mishra (1990) studied hydrobiological characteristics of Morar river in relation to plankton, productivity and pollution. Bharati and Krishnamurthy (1990) showed the effect of industrial effluents on river Kali around Dandeli in Karnataka. Joy *et. al.* (1990) observed the effects of industrial discharges on the ecology of phytoplankton production in the river Periyar (India). Singh and Singh (1990) made observations on pollution studies on river Subernarekha around industrial belt of Ranchi (Bihar). Bharati and Krishnamurthy (1990) observed effects of industrial effluents on river Kali around Dandeli, Karnataka. Rao *et. al.* (1990) studied Ganga water quality in Bihar. Mittal and Sengar (1990) studied on the assessment of water pollution and distribution of algal flora of Karwan river at Agra.

Patralekh (1991) studied periodicity of phytoplankton in the river Ganga at Bhagalpur, Bihar. Birasal *et. al.* (1991) studied the physico-chemical status of Kali river after the closure of the Supa Dam. Chauhan (1991) observed effects of distillery effluents on river Wainganga. Choudhary (1991) studied sediments characteristics of river Ganga in Sultanganj, Bhagalpur region, Bihar. Tiwari *et. al.* (1991) studied the effect of Monitoring the water pollution in snow fed river Alaknanda at Rudraprayag, Chamoli. Tamot and Khare (1991) observed limnological studies of upper lake Bhopal during certain religious activities. Shaji and Patel (1991) studied chemical and biological evaluation of pollution in the river Sabarmati at Ahmedabad (Gujarat). Srivastava (1991) has made observations on pollution study of natural waters of 3 rivers at Gorakhpur.

Mishra and Saksena (1991) studied pollutional ecology with reference to physico-chemical characteristics of Morar (Kalpi) river of Gwalior. Unni *et. al.* (1992) studied preliminary hydrobiological studies of river Narmada from Amarkantak to Jabalpur. Das *et. al.* (1992) studied physico-chemical characteristics of Brahmaputra water at Tezpur. Datar and Vashishtha (1992) studied physico chemical aspects of pollution in river Betwa. Israili (1992) studied occurrence of heavy metals in Ganga river water and sediments. Tripathi and Vencobacher (1992) studied pollution monitoring of Pandu river using remote sensing techniques. Shivakumar *et. al.* (1992) studied impact of viscose rayon factory effluents on the water quality of river Bhavani. Kartha and Rao (1992) studied environmental status of Gandhisagar reservoir (M.P.). Athappan *et. al.* (1992) studied on the pollution of river Vaigal at Madurai (T.N.) Bist and Chopra (1992) studied on seasonal diurnal rhythm of some physico-chemical qualities of river Pinder of Garhwal Himalaya. Pandey *et. al.* (1992) observed seasonal rhythm in the physico-chemical properties of Mahananda river (Katihar, Bihar). Bhatt and Pathak (1992) studied assessment of water quality and aspects of pollution in a stretch river Gomati. Khatvakar and Trivedy (1992) analyzed water quality parameters of river Panchaganga near Kolhapur and Ichalkaranji in Maharashtra. Mishra (1993) studied phytoplanktonica composition of sewage polluted Morar (Kalpi) river of Gwalior (M.P.). Shrivastava *et. al.* (1993) studied pollution of the river Ganga in the Mirzapur region (India) and its impact on sediments. Bilgrami *et. al.* (1993) studied use of quantitative and qualitative indices for evaluating water quality of the Ganga. Chauhan and Pathak (1993) observed Impact of waste waters on water quality of Bihar river, of Rewa. Singh *et. al.* (1993) studied on assessment of BOD load in river Yamuna in respect of potable water of Agra. Ashutosh *et. al.* (1993) observed physico-chemical characteristics of sewage and its impact on water quality of Alaknanda at Srinagar (Garhwal). Ghosh and Kumar (1993) studied impact on surface water due to the discharge of coal washery effluent and dispersion profile of pollutants in Damodar river. Haniffa *et. al.* (1993) studied physico-chemical and microbiological studies in the perennial river Tambaraparani for the assessment of water quality.

Joshi *et. al.* (1993) studied on the physico-chemical characteristics of river Bhagirathi in uplands of Garhwal Himalaya. Balasubramanian (1993) studied water quality predictions in the river Bhavani. Saxena and Chauhan (1993) studied physico chemical aspects of pollution in river Yamuna at Agra. Somasekhara (1993) assessed the quality of Godavari river water during and after the 1991 Pushkaram at Rajahmundry.

Krishnan *et. al.* (1994) studied faecal pollution in river Vaigai (T.N.), Hosetti *et. al.* (1994) studied water quality in Jayanthi Nalla and Panchaganga river at Kolhapur. Singh and Singh (1994) observed variation and correlation of dissolved oxygen with effluent quality and stage of river Ganga at Varanasi (India).

Shrivastava and Singh (1995) observed algal flora in relation to industrial pollution of Rapti river at Gorakhpur. Sahu *et. al.* (1995) studied of some physico-chemical characteristics of the Ganga river water (Rishikesh-Kanpur) within twenty four hours during winter 1994. Mitra (1995) studied water quality of some tributaries of Mahanadi river. Singh (1995) studied pollution status of Rapti river at Gorakhpur. Kumar (1995) studied on pollution of river Mayurakshi in South Bihar.

CONCLUSION

Thus, there are some agents that lie behind the

"Beach closed for swimming
River water unfit for drinking
and needed but never posted."

Today society interest is centered on the effects of pollution on living organisms and therefore the assessment of pollution of Indian rivers is needed. For that the literature of water pollution of rivers with relation to algae is useful for solving the problems.

ACKNOWLEDGEMENTS

The authors are thankful to Principal B.M. Patil S.S.V.P.S.'s L.K. Dr. P.R. Ghogrey Science College, Dhule. We sincerely express thank to different library of Indian Institutes and Universities.

REFERENCES

Adebisi A.A. (1981), The physico-chemical hydrobiology of tropical seasonal river upper Ogun river. *Hydrobiologia* **79**(2):157-166.

Agarwal S.K. (1983), Water quality of sewage drains entering Chambal river at Kota. *Acta. Ecol.* **5**: 29-29.

Allen W.E. (1920), A quantitative and statistical study of the plankton of the Son Joaguin river and Tributaries in and near Stockton California in 1913, *Pub li Zool.* **22**:1-297.

Athappan, P.P. Sethuraman, K; Kannan N. (1992), Study on the pollution of river Vaigai at Madurai. *Indian J. Environ. Prot.* 818-823.

Aykul G. (1978), A quantitative study of the phytoplankton of the river Avon. *Bristol Br. Phykal J.* **13**: p1-102.

Barhate, V.P. and Tarar J.L. (1981), The algal flora of Tapti river, Bhusawal, Maharashtra. *Phykos* **20**: 75-78.

Bharati, S.G. and Krishna M.S.R. Murthy (1990), Effect of Industrial Effluents of River Kali Around Dandeli, Karnataka. *Indian J. Environ. Hlth* **32**(2): 165-171.

Bhatt, S.D. and Pathak, J.K. (1992), Assessment of water quality and aspects of pollution in a stretch of river Gomati (Kumaun) lesser Himalaya *J. Environ Biol.* **13**(2): 113-120.

Bhimcher B.S. and Devid A. (1946), A study of the effects of factory effluents on the Bhadra river, *Fisheries at Bhadrawati 33rd Indian Science Congress Proc.* **3**:310.

Bilgrami, K.S. Kumar, Sheo, Sahy, S.S. (1993), Use of quantitative and qualitative indices for evaluating water quality of the Ganga. *Proc. Indian Nati. Sci. Acad.* **59**B(1): 59-65.

Bist, K.L. and Chopra, A.K. (1992), Studies on seasonal diurnal rhythm of some physico-chemical quantities of river Pinder of Garhwal, Himalaya. *Himalayan J. Env. Zoo* **6**(2): 172-175.

Biswas K. (1942), The role of algae communities of the river Hughly in the drinking water of Calcutta. *150th Ann. Vol. Roy Bot. Gdn.* 189-206.

Blum J.L. (1956), The ecology of river algae. *Bot. Rev.* **22**(5): 291-341.

Brooker M.P. and Johnson P.E. (1984), The behaviour of Phosphate, Nitrate, Chloride and Hardness in twelve Welsh rivers. *Water Res. Vol.* **18**(9): 1155-1164.

Bulusu K.R., Arora H.C., and Aboo K.M. (1967), Certain observations of self purification of river and its effect on Kshipra river. *Enviro. Hlth.* **8**: 275-295.

Butcher R.W. (1947), Studies in the ecology of rivers and the algae of organically enriched waters, VII *J. Ecol.* **35**: 186-191.

Chacko P.I. and Ganapati S.V. (1952), Hydrobiological survey fo Suruli river in the highways, Madurai district India to determining it's suitability for the introduction of Rainbow trout, *Arch. Fur. Hydrobiologie* **66**: 128-141.

Chacko P.I., Shrinivasan R. and Evaneline (1953), Hydrobiology of Malampurzha river, *Ind. Com. J. Madras* 8(1): 118-124.

Chakrabarty R.D., P. Roy and Singh S.B. (1957), A qualitative study of the plankton and the physico-chemical conditions of the river Jamuna at Allahabad in 1954-55. *Indian J. Fish* **6**(1): 186-203.

Chaudhary S.K. (1991), Sediment characteristics of the river Ganga in Sultanganj-Bhagalpur region, Bihar. *Env. Eco.* **9**(1): 249-251.

Chauhan Anil (1991), Effects of distillery effluents on river Wainganga. *Indian J. Environ. Hlth.* **33**(2): 203-207.

Chauhan Ugamkumari and Pathak Dileep (1993), Impact of waste waters on water quality of Bihar river, Rewa, *Proc. Acad. Environ. Bio.* **2**(1): 41-43.

Chopra, A.K.; Madhwal, B.P. and Singh H.R. (1990), Abiotic variables and primary productivity of River Yamuna at Naugaon Uttarkashi, Garhwal (1990). *Indian J. Ecol.* **17**(1): 61-64.

Classen P.W. (1927), Biological studies of polluted areas in Genesee river system, *New York State Ann Rept. for 1992.*

Das H.B., Kalita H., Saikial B., Borah K., Kunnur S.B. (1992), Physico-chemical characteristics of Brahmaputra water at Tezpur. *Polln. Res.* **11**(3): 169-172.

Das, S.M. and V.K. Srivastava (1956), Qualitative studies on fresh water. Plankton. Plankton of fish tank in Lucknow, India. *Proc. Nat. Acad. Sci. (India)* **26**: 8-62.

Das S.M. and Srivastava V.K. (1959), Quantitative studies on freshwater plankton III Qualitative composition and seasonal fluctuations. *Sci. (India)* **29**: 174-189.

Datar M.O. and Vashistha R.P. (1992), Physico-chemical aspects of pollution in river Betwa. *Indian J. Environ. Prot.* **12**(8): 577-580.

David, A. (1963), Studies on fish and fisheries of the Godavari and Krishna river system Part-I, *Proc. Nat. Acad. Sc. Indian* **33B**: 263-286.

De. Smet, W.H.A. Das, E.A. Van. Rompu and J.A. De. Greef (1987), Hydrobiological studies of surface waters in Artwerp (Belgium) II. The water of the moat of fort Merksen *Nat. Ourwet Tijdschr* **69**(4-6): 156-231.

Fjerdingstad (1950), The microflora of river Mollea with special reference to the relation of benthal algal population *Folialimnologica condinonca* **5**: 1-123.

Fritsch F.E. (1903), Observations on phytoplankton of the river Tames *Ann. Bot.* **17**: 631-647.

Ganpati S.V. and Alikunhi K.H. (1950), Factory effluents from Mettur Chemical and Industrial Corporation Ltd. Mettur dam Madras and their effect on the fisheries of the river Cauvery. *Proc. Nat. Inst. Sci. India* **16**: 189-197.

Ganpati, S.V. and Chacko P.I. (1951), An investigation of the river Godavari and the effect of the paper mills pollution at Rajahmundry. *Proce of Indo. Pacific Fish Council* 1-5.

Gunale V.R. and Balkrishnan M.S. (1981), Biomonitoring of Eutrophication in the Pavana, Mula and Mutha rivers flowing through Poona. *Indian J. Environ. Hlth* **23**(4): 316-322.

Gautam Ashutosh, Joshi V.P., Sati O.P. (1993), Physico-chemical characteristics of sewage and its impact on water quality of Alaknanda and Srinagar (Garhwal) *J. Ecotoxico Environ. Monit* **3**(1): 61-63.

Ghosh M.K., Kumar Ashok (1993), Impact on surface water due to the discharge of coal washery effluent and dispersion profile of pollutants in Damodar river *Assian Env.* **15**(1): 32-40.

Ghousuddin, M. (1934), Notes on the algal flora of river Moosi Hyderabad, *Deccan J. Osm. Univ.* 1-3.

Govindan and Sudresan (1979), Seasonal Succession of algal flora in polluted region of Adyar river *Indian J. Environ. Hlth.* **21**(2): 131-142.

Haniffa M.A., Arockiasamy S., Martin P. (1993), Physico-Chemical and microbiological studies in the perennial river Tambaraparani for the assessment of water quality. *Indian J. Environ. Prot.* **13**(7): 533-538.

Holmes, N.T.H. and Whitton B.A. (1981), Phytoplankton of 4 rivers, The Tyne, Wear, Teesand and Swale, *Hydrobiologia,* **80**: 111-127.

Hosetti B.B.; Kulkarni A.R.; Patil H.S. (1994), Water quality in Jayanthi Nalla and Panchaganga at Kolhapur. *Indian J. Environ. Hlth.* **36**(2): 124-127.

Hussan A; Al-Saadt (1996), Algal Ecology and Composition in the Garmat Ali River Iraq. *Regulated Rivers Research and Management* **12**: 27-38.

Irmer V. (1992), Biological warning system for the detection of pollutant incidents of Rhine river. *Schrifter Ver Wasser Boden Lufthya* **89**: 203-210.

Israli Abdul Wahid (1992), Occurrence of heavy metals in Ganga river water and sediments. *Indian J. Environ. Hlth.* **34**(1): 63-66.

Iyengar M.O.P. and Venkataraman G. (1951), The ecology and seasonal succession of the river Cooum at Madras with special reference to Diatomataceae. *J. Madras Univ.* **21**: 140-192.

Jagdale M.H., Salunke M.M., Yabhute Y.B., Kodaskar S.N. and Ghuge K.N. (1984), Pollution of Godavari river water at Nanded *Poll. Res.* **3**: 83-84.

Jeejibai, N. and Rajendra, S. (1980), Phytoplankton Constituents as indicators of water quality. A study of Adyar river *Proc. National Workshop Algal Systems*, 267-270.

Joshi B.D., Pathak J.K., Singh Y.N., Bisht R.C.S., Joshi P.C. (1993), On the physico-chemical characteristics of river Bhagirathi in the uplands of Garhwal Himalaya. *Himalayan J. Env. Zoo.* **7**(1): 64-75.

Joy C.M. Balakrishnan K.P. Ammini Joseph (1990), Effect of industrial discharges on the Ecology of phytoplankton production in the river Periyar (India) *Wat. Res.* **24**(6): 787-796.

Kar G.K., Mishra P.C., Dash M.C. and Das R.C. (1987), Pollution studies in River LbIII; Plankton population and primary productivity. *Indian J. Environ Hlth.* **29**(4): 322-329.

Kartha K.N., Rao K.S. (1992), Environmental studies of Gandhisagar reservior. *Fishery Techno* **29**(1): 14-20.

Khatvakar S.S. and Trivedy R.K. (1992), Water quality parameters of river Panchganga near Kolhapur and Ichalkaranji, Maharashtra, (India). *J. Ecotoxico Environ. Monit* **2**(2): 113-118.

Klein L. (1957), Aspects of river pollution *Butteworth Scientific Publication London.*

Kofoid, C.A. (1903), The plankton of the Illinois river, 1894-1899 with introductory notes upon the hydrography of the Illinois river and its basin. Part-I Quantitative investigations and general results. *Bull. Illinois State Lab. Nat. Hist.* **6**: 95-629.

Krishnan N. Jeyachandran G. and Jeya Kumar (1994), Faecal pollution in river Vaigai. *Indian J. Environ. Hlth.* **36**(2): 128-129.

Lack J.J. (1971), Qualitative studies on the phytoplankton of the river Thames and Kenet at Feading. *Fresh Water Biol.* **1**: 213-224.

Lackey J.B. and Hupp E.R. (1956), Plankton population in Indian's White river. *J. Amer. Water Works Assn.* **48**: 1024-1036.

Lakshminarayan, J.S.S. (1965), Studies on the Phytoplankton of river Ganga, Varanasi India Part-I, the Physico characteristics of river Ganga. *Hydrobiologia* **25**: 119-137.

Lakshminarayan J.S. (1965-b), Studies on phytoplankton of river Ganga, Varanasi Part-IV phytoplankton in relation to fish pollution *Hydrobiologia* **25**: 171-175.

Lal, A.K. Sarkar, A. Sarkar, S.K. and Shah, K.L. (1982), Species diversity of periphyton as on index of population of river Ganga. *J. Int. Fish. Soc. Ind.* **14**: 44-51.

Lam, C.W.Y. (1971), Dynamics of phytoplankton growth in the Waiketo river North New-Zealand. *Hydrobiologia,* **66**(3): 237-244.

Mahajan K.K. (1988), Deteriorating Nations river In: "*Ecology and pollution of Indian rivers*", (R.K. Trivedy et.) pp. 1-38.

Mishra S.R. (1980), Hydrobiological characteristics of Morar river Gwalior in relation to plankton and productivity. *Ph.D. Thesis Jiwaji Univ., Gwalior (M.P.)*

Mishra S.R. and Saksena D.N. (1992), "*Aquatic Ecology*" Ashish Publishing House, New Delhi-110026.

Mishra S.R. (1993), Phytoplankton composition of sewage polluted Morar (Kalpi) river in Gwalior (M.P.) *En. Eco.* **11**(3): 625-629.

Mitra A.K. (1995), Water quality of some tributaries of Mahanadi. *Indian J. Environ. Hlth.* **37**(1): 26-36.

Mittal Sharad, Sengar R.M.S. (1990), Studies on the assessment of water pollution in Karwan river. *Poll. Res.* **9**: 91-94.

Motwani M.P., S. Banerjea and Kamchandani S.J. (1956), Some observations on the pollution of river Sone in the factory effluents of the Rohtas industries at Dalmianagar (Bihar) *Indian J. Fish* **3**:334-367.

Nandan S.N. and Patel R.J. (1983), Study of Algal communities in river Vishwamitri, Baroda as indicators of organic pollution *Indian J. Ecol.* **10**(1): 11-15.

Nandan S.N. and Patel R.J. (1984), Ecological studies on algal flora of Vishwamitri river Baroda, Gujarat, *Indian J.P.I. Nature* **2**(1): 17-22.

Nandan S.N. and Patel R.J. (1985), Eutrophication in Vishwamitri river flowing through Baroda city *Geobios* **2**: 60-62.

Nandan S.N. and Patel R.J. (1986), Assessment of water quality of Vishwamitri river by algal analysis. *Indian J. Environ Hlth.* **29**(2): 160-161.

Nandan S.N. and Patel R.J. (1990), Hydrobiological study of Vishwamitri river, Baroda proceeding, *Academy of Environmental Biology Environmental Impact on Biosystems* 189-193.

Nandan S.N. and Patel R.J. (1992), Ecological studies of algae *Aquatic Ecology*, 69-100.

Nautiyal Prakash Pokhriyal R.C.; Gautam A.; Rawat D.S. and Singh H.R. (1988), Maneri dam on the river Bhagirathi. *A lacustrine Environment in the making Proc. Nat. Symp. Past, Present and Future of Bhopal Lakes* 175-182.

Palmer C.M. (1959), Algae in water supplies U.S. Dept Hlth Efuc., Welf Pub. Hlth. Serv. Bur Stage Service Div. Water Pollution control. Robt. *A. Taff. San. Engr. Center, Cinnanati. Ohio. Pub. Hlth. Serv. Publ. No.* 675-88P.

Palharya J.P. and Malvia S. (1988), Pollution of the Narmada river at Hoshangabad in Madhya Pradesh and suggested measures for control. In: *Ecology and Pollution of Indian rivers* (R.K. Trivedy Ed.) pp. 54-85. Ashish Publishing House New Delhi, India.

Pandey B.N., Lal R.N., Mishra P.K. (1992), Seasonal rhythm in the physico-chemical properties of Mahananda river. Katihar Bihar. *Env. Eco.* **10**(2): 354-357.

Paramasivam M. and Sreenivasan A. (1982), Distribution of Diatoms in the Cauvery river system with special reference to pollution. *J. Environ. Biol.* **3**:1-9.

Patra A.K. and Nayak I.D. (1982), Limno biotic survey of river Mahanadi during winter. *Geobios New Reports* **1**:20-23.

Patralekh L.N. (1951), Periodicity of phyto-plankton in the river Ganga at Bhagalpur Bihar, India. *Env. Eco.* **9**(1): 84-86.

Pingale S.D. (1981), Studies on the algae of impoundments and streams in Maharashtra Ph.D. Thesis Poona University.

Prasad B.N. and Singh Yashpal (1980), Algal Hydrobiology in India, *A Review Nat. Aca. Sci. India Golden Jubilee Commemoration* Vol. 271-300.

Prasad B.N. and Saxena M. (1980), Ecological study of blue-green algae in river Gomati. *Ind. J. Environ. Hlth* **22**: 151-166.

Prasad, B.N. and Sing Y. (1982), Species diversity of benthic diatoms in lotic environment. *Indian J. Environ. Hlth* **24**: 308-315.

Quasium, S.Z. and Sidiqui R.H. (1960), Preliminary observations of river Kali covered by the effluents of industrial wastes. *Current Science,* **29**(8): 310-311.

Rai H. (1962), Hydrobiology of river Yamuna at Okhla, Delhi, India, Scientific paper from Inst. of Chemical *Technology Prague Technology of water* **6**: 77-79.

Rai H. (1974 a), Limnological studies on the river Yamuna at Delhi, India Part-I. Relation between the chemistry and the state of pollution in the river Yamuna *Arch. Hydrobiol.* **73**(3): 369-393.

Rai H. (1974 b), Limnological studies on the river Yamuna at Delhi, India Part-II. The dynamics of potamoplankton population in river Yamuna. *Arch. Hydrobiol.* **73**: 492-517.

Rai L.C. (1978), Ecological Studies on algal communities of River Ganga at Varanasi. *Ind. J. Ecol.* **5**: 1-6.

Raina V., Shah A.R. and Ahmed S.R. (1984), Pollution studies on river Jhelum: An assessment of water quality, *Indian J. Environ. Hlth.* **26**: 187-201.

Ramarap, S.V., Singh V.P., and Mali L.P. (1978): Pollution studies of river Khan (Indore). India-I Biological assessment of pollution. *Water Research,* **12**: 555-559.

Rana B.C. and Palgia S. (1988), Assessment, evaluation and abstatement studies of Polluted river Bandi and (Rajasthan). In *Ecology and Pollution of Indian Rivers.* (R.K. Trivedy et) pp. 345-359. Ashish Publishing House, New Delhi.

Rao, S.N., Chaubey R. and Shrinivasan K.V. (1990), Ganga water quality in Bihar. *Indian J. Environ. Hlth.* **32**(4): 393-400.

Reinhard E.G. (1931), The plankton ecology of the upper Misissippi Minneapolis to Winona *Eco. Monagraphs* **1**: 396-469.

Rice C.H. (1938), Studies in the phyto-plankton of river Thames I and II. *Ann. Bot. N.S.* **2**: 539-577.

Rishi, V. and Kachroo, P. (1981), Effect of sewage and fertilizers on phytoplankton of Doodhganga *Geobios* **8**: 160-162.

Roy H.K. (1955), Plankton ecology of river Hooghly at Patana, West Bengal. *Ecology* **36**: 169-175.

Rzoska (1976), *The Nile biology of an ancient river monographic.* Monograph. Biol. **29**: 1-417.

Saha L.C., Choudhary S. and Singh M.N. (1985), Factors affecting phytoplankton productivity and density in the river Ganga at Bhagalpur. *Geobios* **12**: 63-65.

Sahu B.K., Rao R.J. and Behera (1995), Studies of some physico-chemical characteristics of the Ganga river water (Rishikesh-Kanpur) within twenty four hours during winter 1994. *Ecol. Env. and Cons.* **1**(1-4): 35-38.

Sankaran, V. (1984), Algae of the Aliyar river. *Phykos* **23**: 75-79.

Saxena K.K., Chauhan R.R.S. (1993), Physico-chemical aspects of pollution in river Yamuna at Agra. *Polln. Res.* **12**(2): 101-104.

Sengupta B., Laskar S., Das A.K. and Das J. (1988), Inorganic pollutants of Ganga water in the region of Berhampore to Ketwa, West Bengal. *Indian J. Environ. Hlth.* **30**(2): 202-208.

Shah A.R. (1988), Physico-chemical aspects of pollution in river Jhelum (Kashmir) during 1981-83. In *Ecology and pollution of Indian rivers* (R.K. Trivedy ed.) 163-207. Ashish Publishing House, New Delhi.

Shaji C., Patel R.J. (1991), Chemical and biological evaluation of pollution in the river Sabarmati at Ahmedabad, Gujarat. *Phykos* **30**(1&2): 91-100.

Shreenivasan, Sampath A.V., Paramasivum M. and Anonthanarayanan (1980), Pollution of river Cauvery from Industrial and Urban wastes. *Proc. Symp. Environ. Trivandrum* 178-190.

Shukla S.C., Tripathi B.D., Kant Rajani, Deepa Kumari V. and Panday V.S. (1989), Physico-chemical and Biological Characteristics of River Ganga from Mirzapur to Ballia. *Indian J. Environ. Hlth.* **31**(3): 218-227.

Shukla S.C., Kant R. and Tripathi B.D. (1989), Ecological investigation on physico-chemical characteristics and phytoplankton productivity of river Ganga at Varanasi. *Geobios* **16**: 20-27.

Singh B.B. (1995), Pollution status of Rapti river at Gorakhpur. *Journal of Environment and Pollution* **2**(3): 117-120.

Singh D.K. and Singh C.P. (1990), An assessment of water quality of river Subernarekha (Bihar) *Geobios* **17**: 171-175.

Singh D.K. and Singh C.P. (1990), Pollution studies on river Subernarekha around industrial belt of Ranchi (Bihar) *Indian J. Environ. Hlth.* **32**(1): 26-33.

Singh G.S., Singh A.S. (1994), Variation and correlation of Dissolved oxygen with effluent quantity and stage of river Ganga at Varanasi (India). *Indian J. Environ. Hlth.* **36**(2): 79-83.

Singh J.P., Singh S.K., Singh Khakun, Yadava P.K., Joshi Madhu (1993), Assessment of BOD load in river Yamuna in respect of potable water of Agra. *Indian J. Environ. Prot.* **13**(2): 119-123.

Sinha M.P. (1988), Effect of waste disposal on water quality of river Damodar in Bihar. Physico-chemical characteristics.

In *Ecology and Pollution of Indian Rivers* (R.K. Trivedy ed) pp. 219-246. Ashish Publishing House, New Delhi, India.

Sivakumar A.A., Lakeshmaneswamy M., Lakshimi R., Iseyvani A. (1992), Impact of viscous Rayon Factory effluents on the water quality of the river Bhavani *Himalayan J. Env. Zoo.* **6**(12): 97-101.

Somashekar R.K. (1984), Studies on water pollution of the river Cauvery, Physico-chemical characteristics. *Intern J. Environ. Studies* **23**: 209-216.

Somasekhara Rao Kaza, Padamavathy D. Rambabu C. (1993), Monitoring the quality of Godavari river water during and after the 1991 Pushkaram at Rajahmundry. *Polln. Res.* **12**(3): 191-195.

Srivastava Ajai, Mehrotra M.N., Tiwari R.N. (1993), Study of pollution of the river Ganga in the Mirzapur (India) and its impact on sediments. *Int. J. Environ Stud.* **43**(2-3): 201-208.

Srivastava G.K., Singh B.B. (1995), Observation of algal flora in relation to industrial pollution of Rafti river at Gorakhpur. *Eco. Env. and Cons.* **1**(1-4): 53-55.

Sudhakar, G. and Venkateshwarlu V. (1989), Ecological imbalances in the river of Andhra Pradesh (India). Effects of paper mill effluents on the river Tungbhadra and Godavari. *Int. J. Environ. Stud.* **34**: 89-97.

Sunder, S. (1988), Monitoring the water quality in a stretch of river Jhelum, Kashmir. In *Ecology and Pollution of Indian Rivers* (R.K. Trivedy ed.) pp. 131-161. Ashish Publishing House, New Delhi, India.

Tamot Praveen, Khare Satyendra, Shrivastava Pradeep (1991), Limnological studies of upper lake Bhopal during certain religious activities. *J. Nature Conserv.* **3**(2): 141-146.

Tiwari N.C., Sagar Gyan, Tiwari Dinesh, Singh H.R. (1991), Monitoring the water pollution in snow fed river Alaknanda at Rudraprayag, Chamoli. *Env. Eco.* **9**(1): 202-206.

Tiwari T.N. (1989), Pollution of river Jhelum in Kashmir : Cluster analysis of its water quality parameters. *Indian J. Envi. and Agri.* **4**: 102-110.

Tripathy P.K. and Adhikary S.P. (1990), Preliminary study on the water pollution of river Nandira. *Indian J. Environ. Hlth.* **32**(4): 363-386.

Tripathi Nitin, Vancobacher C. (1992), Pollution monitoring of Pandu river using remote sensing techniques. *NNRMS BULL.* 36-38.

Unni K.S., Chauhan A., Varghese M. and Naik L.P. (1992), Preliminary hydrobiological study of river Narmada from Amarkantak to Jabalpur. In "*Aquatic Ecology*" (Mishra S.R. Saksena, D.N. Eds.) pp. 221-229. Ashish Publishing House, New Delhi.

Vass K.K., Raina H.S., Zutschi O.P. and Khan M.A. (1977), Hydrobiological studies on river Jhelum *Geobios* **4**: 238-242.

Vanlandigham S.L. (1976), Comparative evaluation of water quality on the St. Joseph river (Michigan and Indiana USA) by 3 methods of algal analysis. *Hydrobiologia* **48**: 145-173.

Verkey J. and Alexander (1968), A preliminary report on hydrobiology of Beypore river for the year 1964, *Hydrobiologia* **31**: 492-496.

Venkateswarlu V. (1969a), An ecological study of algae of the river Moosi, Hydrobiol. (India) with special reference to water pollution-I. Physico-chemical complexes. *Hydrobiologia* **33**: 117-143.

Venkateswarlu V. (1969b), An ecological study of algae of river Moosi, Hyderabad (India) with special reference to water pollution-II. Factors influencing the distribution of Algae *Hydrobiologia* **33**: 352-363.

Venkateswarlu V. (1969c), An ecological study of algae of the river Moosi, Hyderabad (India) with special reference to water pollution-III. The algal periodicity *Hydrobiologia* **34**: 533-560.

Venkateswarlu V. and P.T. Sampat Kumar (1982), Chemical and Biological assessment of pollution in the river Moosi Hyderabad (A.P.) India. *Biol. Bull India* **4**: 23-30.

Venkateswarlu V. and Reddy Manikya (1985), Algae as Biomonitors in river ecology. *Sym. Bio. Monitoring State Envir.* 183-189.

Verma S.R. and Delela R.C. (1975), Studies on the pollution of the Kalinadi by industrial wastes near Mansurpur Part-I : hydrometric and physico-chemical characteristics of the wastes and river water. *Acta. Hydrochim, Hydrobiol.*, **3**: 239-275.

Wasylim (1965), Communities of algae from the Sola river and its tributaries *Actahydrobiol* **7**: 9-60.

Weber C.I. and Moore D.R. (1967), Phytoplankton seston and dissolved organic carbon in little Miami river at Cincinnati. *Chio-limnolo. Oceanog.* **12**: 311-318.

Willmen, H.O., Desment M.J., Frans C. Evens (1972), A hydrobiologican study of the polluted river Lieve (ghent-Belgium) *Hydrobiologia* **39**(1): 91-154.

Yoshio O. and Ichinura (1984), Phytoplankton diversity in inland water of different trophic status. *J. Limnol.* **45**(3): 137-177.

Zingde M.D., Marvekar P.V., Sharma R.V., Desai B.N. (1980), Water quality of the river Damanganga (Gujarat). *Indian J. Marine Sci.* **9**: 94-99.

3

Tendencies in Water Quality Research and its Application to Twenty First Century

T. Venkatesan, P. Geetha, G. Padmalatha, S. Balasubramanian & B. Suresh

INTRODUCTION

Most of the water quality models ignore the fact that natural systems are inherently – stochastic and hence they are of limited value to decision makers. The Biochemical Oxygen Demand (BOD) and Dissolved Oxygen Level (DO) models originate from the pioneering water quality studios of Streeter-Phelps. The basic assumption of this model is that bio-degradable wastes, which are discharged into a river system exerts oxygen demand and the atmospheric re-aeration is the only source of Oxygen (Gromiec, 1983). The non-conservative pollutants are reduced in quantity over time and changed in character by the physical, biological and chemical processes that are exerted in the river system. The conservative substances are not degraded with time, but only by their dilution. Therefore, separate models are described to account the BOD and DO levels in a river. Streeter & Phelps (1925) have developed a model to establish the relationship between BOD–DO in a stream system with the assumption of (a) A constant load of pollutants are discharged at a given point, (b) the cross-section of the river and the turbulence is sufficiently intensive, (c) the water and the pollutants are mixing perfectly (a first-order ordinary differential equation with transport delay between spatial points), partially (a second-order partial differential equation) and no mixing (a first-order partial differential equation) and (d) the downstream flow of water is constant.

THE MODEL

Relationship of DO and BOD are based on the re-aeration and de-oxygenation (oxygen sag) in the liquid-air interface, which shows a first-order kinetic reaction. The differential equations explaining the de-oxygenation and re-aeration rates are :

$$\frac{dD}{dt} = -\frac{dL}{dt} = K_d L \qquad ...(1)$$

and

$$\frac{dD}{dt} = -K_a D \qquad ...(2)$$

where

D = the saturation deficit, or the difference between the DO saturation concentration and the actual DO concentration (MI^{-3})

L = the carbonaceous biochemical oxygen demand, (CBOD) (MI^{-3})

K_d = the de-oxygenation coefficient (T^{-1})

K_a = the re-aeration coefficient (T^{-1})

t = time of flow (T)

The balance amount of DO is calculated from equation 1 and 2 as

$$\frac{dD}{dt} = K_d L - K_a D \qquad ...(3)$$

The exact solution for equation (3) is :

$$D_t = \frac{K_d L_o}{K_a K_d} \exp(-K_d t) - \exp(-K_a t) + D_O \exp(-K_a t) \qquad ...(4)$$

where

D_t = the dissolved oxygen saturation deficit after time of flow t (MI^{-3})

L_o = the oxygen demand at an initial reference point (t = 0) in the river (MI^{-3})

D_o = the dissolved oxygen deficit at the same point (MI^{-3})

K_d = the de-oxygenation rate coefficient (T^{-1})

K_a = the re-aeration rate coefficient (T^{-1}) (Gromiec 1983)

Equation 4 is the oxygen sag formula of Streeter-Phelps and a fitted curve with time in the X axis and DO level in the Y axis is known as oxygen-sag curve (Arbabi *et al* 1974; Young and Beck, 1974; Jorgensen, 1983; Lamp, 1985).

The basic Streeter-Phelps model still dominates and were used for different purposes to calculate the rate of self-purification of a non-tidal river system.

The self-purification factor (f) = $\frac{K_a}{K_d}$ (Fair, 1939; Fair and Geyer, 1965). The DO deficit (D_C) and at time (t_C) defined by the following equations :

$$D_C = \frac{L_O\, 10^{-kdtc}}{f} \qquad ...(5)$$

$$t_C = \frac{1}{K(d)\,(f-1)} \log \left[f - (f-1)\frac{D_O}{L_O}\right] \qquad ...(6)$$

Equations 5 and 6 are used in analysing the influence of the parameters which interfere the re-aeration rate of the stream.

MODIFICATION OF THE STREETER-PHELPS MODEL

The classical BOD and DO models has since then be improved in various ways by other investigators to account a particular process occurring in the river system was reviewed by Gromiec (1983) based on the nature of the pollutants. This will explicitly revealed the increased awareness of the water pollution problem and thus the continuous change in the modelling approach by Scientists and the International organizations.

Thomas (1948) introduced a term K_d in the equation 1, which emphasise the rate constant of BOD removal by sedimentation (T^{-1}). Another variable to be considered in BOD and DO model is the longitudinal dispersion coefficient of the pollutant ($L^2\, T^{-1}$). However the quality parameters do not vary with the temporal mean depth (Camp, 1963).

However, when modelling the BOD and DO in estuaries, the above variable is insignificant, only organic matter oxidation and atmospheric re-aeration are assumed to occur (O'Connor, 1960). Li (1962) considered that the effects of addition of discharge and pollutant along the river, which ultimately influence the fluctuations of the BOD and DO with time at the outfall section. Later in 1964, Dobbins revealed that in addition to the above parameters, the cross-sectional area of the stretch is also to be considered with time.

Models of Liebman (1965) and Liebman and Lynn (1966) emphasized that the saturation level varies linearly with time (within a reach), in order to provide continuity of the DO profile from reach to reach. Frankel (1965) incorporated a photosynthetic effect into a diurnal dissolved oxygen profiles by utilizing a time-dependent factor F′ (Σ). The latter development of O'Connor's (1967) equation to generalize for the temporal and spatial distribution of conservative or non-conservative substances in a one-dimensional river system with the assumption of initial and arbitrary conditions. He assumed that the temporal form of photosynthetic oxygen source may be described with periodic function by a Fourier series. As explained in the foregoing survey of various BOD and DO models, it is possible to predict the de-oxygenation and re-aeration constant from a point source of pollution downstream in different distances with time (Deacon and Giglio, 1971).

The above models are mainly based on first and second order differential equations. But, in the river pollution analysis, there is a possibility of non-point source (diffuse) of pollution. In this case, now the model became a non-linear one. The non-linearity of the environmental systems with integral functions do not have a closed solution. To solve such problems the natural system can be decomposed into a parallel number of linear subsystems of higher degree or higher order (polynomial). Such a solution can be applied into a number of systems like hydrology, air pollution, river channel morphology, BOD levels in river and climatic regions etc. (Bennett and Chorley, 1978).

Bennett *et al* (1976) have analysed the changing persistence pattern of pollution concentration of SO_2 and smoke in London. This non-linear response of the different time periods is expressed in modified autocorrelation and spectral functions with increased annual and monthly cyclic components with evenly spaced discrete times. For example O' Connor (1967) and Yeujevich (1972) discussed this periodicities and trends in hydrological data accounting the annual discharge in a stream ($\Delta t = 1$ year), monthly average inflow/outflow ($\Delta t = 1$ month) and daily rainfall ($\Delta t = 1$ day).

As emphasized by various authors in the foregoing accounts that system analysis in modelling natural systems is closely related to input-output parameters. Hence the environmental/water quality controls is mainly to predict the output system based on the input. We have studied certain of such systems, where the input of the effluent affects the DO level in the river Bhavani through mathematical models :

1. M. Chandra Shekar *et al* 1991 Statistical studies on the correlation of dissolved oxygen levels with Environmental factors in Amaravathi river (South India).
2. M. Chandra Sekar and Balasubramanian S. 1992 Study on oxygen sag curve prediction and forecasting,
3. S. Rajagopal *et al* 1993 water quality predictions in the river Bhavani (South India), and
4. N. Maragatham *et al* 1996 Stochastic analysis of hydrological data of Amaravathi river basin in India.

The above models were used for statistical extrapolation to predict dissolved oxygen concentrations in streams subject to organic waste loads. But, this appıoach is not sufficient to represent the complex ecological interactions involved in the water quality in a stream. Because of the limitations of the existing methods prompted a search for an improved analytical technique for incorporating stochasticity in to environmental models. O' Connor and Di Toro (1970) developed a DIURNAL model to evaluate and predict diurnal variations of DO from photosynthetic activities of aquatic plants. They extended the periodic function of the photosynthetic activities, into a Fourier series, where the photosynthetic source is assumed for a long time to produce a periodic steady-state condition. Later in 1983, Deb and Bowers applied this technique to stimulate the diurnal DO effects in South Fork of the Shenadosh river.

Hence the internal structure of any time series models seems to have various theoretical and practical implications. The degree of interdependence within the successive values (correlation) of time series can be used in detecting the pattern of movement as well as in constructing stimulatory models Shahane *et al* (1977).

Such a correlation analysis (serial) is applied in the present study by Padmalatha (1991) for the DO and BOD data, which is solely related to serial correlation coefficient (autocorrelation).

The variations in the r_k values in different stations indicate many mathematical properties like periodicity, presence of Markov dependence and degree of randomness. This reveals that the water quality of the river Bhavani varies in relation to the quantity of the input of Wasteloads in the river in different months of a year, provided, we assume that all other variables which are influencing the BOD and DO levels are constant in the stretch studied. But there is yet another possibility of the change in the BOD – DO levels in the stream apart from the Wasteloads is the upstream discharge (level) into the river.

The study clearly indicates that the same value is not repeated adjacently, then we can say that the occurrence of an event in the time serial series relates more to its second-ranked sequential event than to its immediate neighbour or to its third ranked event.

We have also used the spectral analysis method for analysing the data. An overall analysis of the periodic characteristic of DO and BOD time series in different stations of river Bhavani reveals variations in frequency and found that the periodicity of the DO and BOD values in the stream again depends on the input wasteloads.

CONCLUSION

Both the models presented here are more advantageous than the other linear correlation techniques. Because, from the models it is easy to observe that, the difference in the spike value of each lags indicates the non-dependability of the preceding's or succeeding lags. However the present study, the available data are very limited. Therefore the spikes and periodicities are not distinguishly established.

REFERENCES

Arbabi M., Elzinga J. and Revelle C. (1974), Water Resources Res., **10**(5): 921-929.

Bennett R.J., (1976), Earth Surface Processes, **1**: 131-150.

Bennett, R.J. and Chorley, R.J. (1978), Environmental Systems, Philosophy, Analysis and Control Methuen Co. Ltd., London.

Camp, T.R. (1963), Water and its Impurities. Reinhold Publishing

Corp. Chapman and Hall Ltd., London.

Chandrashekar M., Rajagopal S. and Balasubramanian S. (1991) Environment & Ecology **9**(1): 77-80.

Chandrashekar M. and Balasubramanian S. (1992), Presented at Enviro Asia–International Conference & Exhibition on the Environment, Singapore.

Deacon, R.W. and Giglio R.J. (1971), Water Resources Centre and Department of Engineering and Operations Research, University of Massachusetts, Amherst, Mass.

Deb, A.K. and Bowers, D. (1983), Journal WPCF, **55**(12): 1476-1488.

Dobbins, W.E. (1964), J. Sani. Eng. Div. Proc. ASCE 90 SA3: 53.

Fair, G.M. (1939), Sewage Works Journal **11**(3): 445.

Fair, G.M. and Geyer, J.C. (1965), Water Supply and Waste Water Disposal, John Wiley and Sons, New York, p. 828.

Frankel, R.J. (1965), An Engineering Research Laboratory Report No. 65-3, University of California, Berkeley.

Gromiec, M.J. (1983), Application of Ecological Modeling in Environmental Management, Part-A, Elsevier Scientific Publishing Company, Amsterdam.

Jorgensen, S.E. (1983), Application of Ecological Modeling in Environmental Management, Part-A, Elsevier Scientific Publishing Company, Amsterdam.

Lamp, C. (1985), Water Quality and its Control, John Wilcy and Sons, Newyork.

Liebman, J.C. (1965), The Optimal Allocation of Stream DO Resources Ph.D. thesis, Cornell University, Ithaca, Newyork.

Liebman, J.C. and Lynn, W.R. (1966), Water Resources Res., **2**(3): 581.

Li (1962), J. Sani. Eng. Div. Proc. ASCE 88 SA3: 75.

Maragatham N., Venkatesan T., Geetha S., Murugesan K. and Balasubramanian S. (1996), Modeling, Measurement & Control 54 N2: 55-64.

O' Connor, D.J. (1960), J. Sani. Eng. Div. Proc. ASCE 86 SA3: 35.

O' Connor, D.J. (1967), Water Resources Res., **3**(1): 65.

O' Connor, D.J. and Di Toro, D.M. (1970), J. Sani. Eng. Div. Proc. ASCE 96 SA2: 547.

Padmalatha, G. (1991), M.Phil. Thesis, Bharathiar University, Coimbatore.

Rajagopal S., Lalithaparameswari N., Padmalatha G., Geetha S. and Balasubramanian S. (1993), Environment & Ecology **11**(3): 534-539.

Shahane A.N., Thomas D. and Bock P. (1977), Water Resources Res., **13**(1): 93-100.

Streeter, H.W. and Phelps E.B. (1925), Public Health Bulletin No. 146, U.S. Public Health Service.

Thomas A.H. (1948), Water and Sewage Works, 95: 409.

Yeujevich, V.M. (1972), Hydrology paper 56, Colorado State University, Fort colline, Colorado.

Young, P.C. and Beck, B. (1974), Automatica. **10**: 455-468.

4

Occurrence and Hygienic Significance of Aeromonads in River Narmada

Anjana Sharma & Sushma Rajput

INTRODUCTION

Aeromonas species are widely distributed in aquatic environments (Hazen *et al*, 1978). These bacteria are commonly observed in both polluted and unpolluted fresh waters, in sewage as well as in drinking water (Burke *et al*, 1983; Monfort and Baleux, 1990; Sharma *et al* 1992).

At one time, *Aeromonas* spp. were thought to cause diseases in immuno-compromised humans, as primary pathogens of fish, reptile and amphibians (Hazen *et al*, 1978, Hird *et al*, 1980). Recently, attention has focused on these micro-organisms because of their increasing association with wound infection exposed during aquatic activity (Beaurne *et al*, 1978; Joseph *et al*, 1979; Thonon *et al*, 1987). Reports from many part of the world suggest that *Aeromonas* are causative agents of acute diarrhoeal diseases in humans following ingestion of contaminated water (Burke *et al*, 1984; Kirov *et al*, 1986; and Mergrand 1986). The presence of these organisms is undesirable because they also interfere with the determination of coliforms bacteria producing false positive reactions (Leclere, 1972). Moreover, *Aeromonas* are known to be an opportunistic pathogens for humans.

The present investigation deals with the occurrence, and nutritional requirements of *Aeromonas* and their hygienic significance in river Narmada at Jabalpur.

MATERIALS & METHODS

Water samples were analysed for physico-chemical characteristics i.e. pH, Temperature, Chloride, Phosphate, Nitrate, Dissolved Oxygen and Bio-chemical Oxygen Demand following standard methods (APHA, 1989).

The density of total heterotrophic bacteria, total coliform and faecal coliforms were made following standard methods (APHA, 1989).

Quantitative estimation of *Aeromonas* were made following membrane filtration technique (APHA, 1989; Rippy & Cabelli, 1985).

Observation on different biochemical characteristics of the isolates were undertaken following Bergey's Manual of determinative bacteriology (9th ed. 1984, Kooij & Hignen, 1988) and with the help of PIB Computer Kit (Bryant, 1989).

An initial screening of the isolates for various nutritional sources was made by replica plating test (Kooij & Hignen, 1988).

The ability of *Aeromonas* to utilise NH_4+N and/or NO_3–N as nitrogen source was determined following periodic measurements at 650 nm. (Kooij & Hignen, 1988).

RESULTS AND DISCUSSION

The physio-chemical and bacteriological studies revealed, the significant seasonal distribution of *Aeromonas* (Table 1 and Fig. 1). River Narmada at it's different sites showed, high counts of *Aeromonas* (upto 189×10^3 CFU/ml) with detectable seasonal pattern, parallel to total coliforms. This indicates that the bacteria at these levels were indigenous to river environment. This may be due to excessive discharge associated with storm water runoff, human and animal waste.

Aeromonas were identified as *A. hydrophila, A. salmonicida, A. sobria, A. salm* and *A. caviae*. During present investigation, we tried to determine the types of compounds which are potential substrate for the growth of *Aeromonas* in fresh water.

A total number of 27 organic compounds were tested as growth substrates in the replica plating test with *A. hydrophila*, which had been isolated from different sites of river Narmada. The objective of this test was to identify organic compounds that could potentially serve as growth substrate.

TABLE 1 : PHYSICO-CHEMICAL AND BACTERIOLOGICAL CHARACTERISTICS OF RIVER NARMADA AT JABALPUR (MEAN AVERAGE VALUES)

Sampling sites	pH (Units)	Temperature (C)	Chloride (mg/l)	Phosphate (mg/l)	Nitrate (mg/l)	D.O. (mg/l)	B.O.D. (mg/l)	HPC (CFU/ ml)	TC (MPN/ 100ml)	FC (MPN/ 100ml)	Aeromonads (CFU/ml) $\times 10^3$
Gwarighat I	7.9	21.0	14.0	22.0	0.17	5.8	3.8	70	1600	900	198
Gwarighat II	7.6	22.0	12.0	27.0	0.26	7.2	4.2	92	900	900	160
Gilahirighat	7.8	22.4	10.8	38.0	0.14	9.2	2.2	89	900	500	189
Tilwaraghat	8.1	26.0	9.8	38.0	0.21	7.8	2.4	46	1600	900	90
Panchvati	8.0	26.0	11.8	32.0	0.19	6.4	3.2	82	900	350	79
Bheraghat	7.8	23.0	12.2	20.0	0.24	8.4	4.4	61	900	280	68

D.O. = Dissolved Oxygen;
B.O.D. = Biological Oxygen Demand;
COD = Chemical Oxygen Demand;
HPC = Heterotrophic plate count;
TC = Total Coliforms;
FC = Faecal Coliforms.

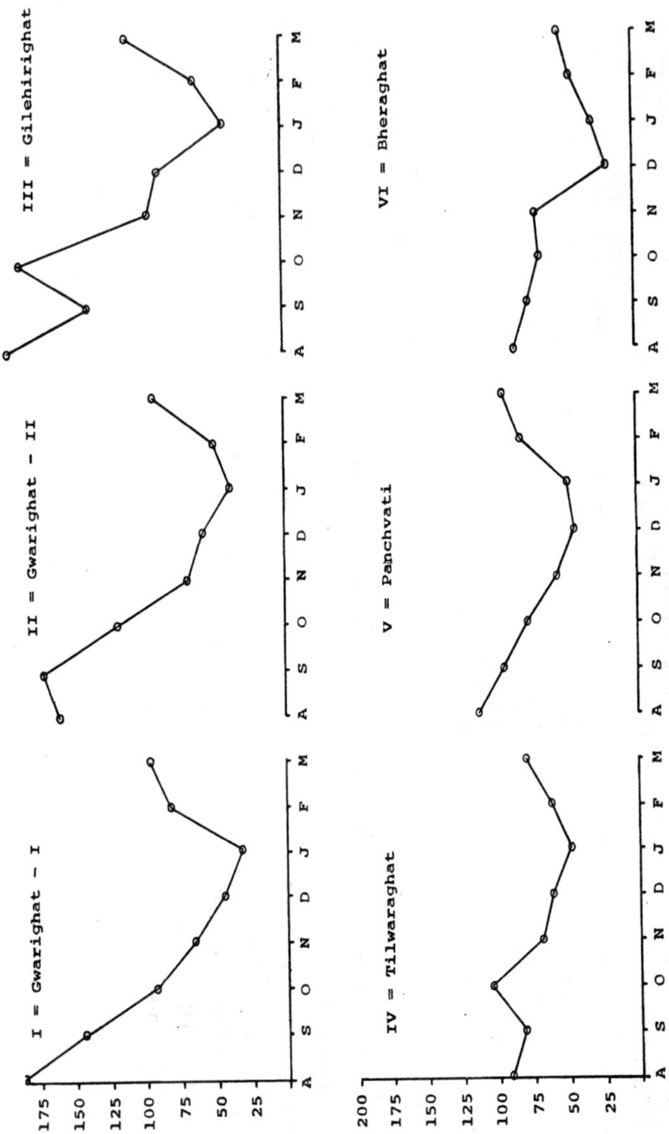

FIG. 1 : SEASONAL DISTRIBUTION OF AEROMONADS IN RIVER NARMADA AT DIFFERENT SITES OF JABALPUR (AUG. 93 TO MARCH 94)

Five amino acids (Arginine, Ornithine, Serine, Aspartate and Glutamate) two carboxylic acids (Citric acid and Succinic acid) five carbohydrates (Galactose, Sucrose, Fructose, Maltose and Arabinose) and one organic alcohol (Mannitol) promoted growth of the isolates. While other amino-acids, alcohols and carbohydrates did not promote growth of the isolate (Table 2). It is concluded from the data obtained that *A. hydrophila* utilized a variety of organic compounds, particularly, amino acids and carbohydrates as the sole source of carbon and energy, when present at a concentration of only few micrograms per litre.

TABLE 2 : COLONY COUNTS AND GROWTH YIELD OF AEROMONADS AT DIFFERENT SITES OF RIVER NARMADA IN JABALPUR

Sampling Sites	Maximum Colony count (CPU/1000 ml)	DOC (mg/l)	No_3- Conc. (mg/l)	pH Unit
1. Gwarighat I	180×10^3	3.9	0.29	7.7
2. Gwarighat II	160×10^3	5.8	0.27	7.8
3. Gilharighat	189×10^3	2.3	0.19	8.0
4. Tilwaraghat	101×10^3	2.1	0.23	8.2
5. Panchvati	110×10^3	2.9	0.14	7.6
6. Bheraghat	80×10^3	2.4	0.12	7.7

The replica plating technique proved to be a valuable tool for the rapid screening of large number of compounds for their ability to serve as the sole source of carbon and energy for the growth of *Aeromonas*. The results obtained with this technique demonstrate that 13 (48%) of 27 organic compounds tested were uniformly utilized by *A. hydrophila* at a concentration of 2.5 g/ litre (Table 3). In comparison to other bacteria like *Pseudomonas aeruginosa* the nutritional versatility of *A. hydrophila* as determined with the replica plating test is more limited in respect to the utilization of amino acids and carboxylic acids (Stainier, 1966; Vender Kooij, 1982; Kooij, 1991).

NH_4+N was more favourable nitrogen source for *A. hydrophila* than was NO_3–N. All the isolates grew in basic salt solution with citrate and NH_4+N (Table 4 and Fig. 2). Thus, the data on nutritional versatility of *A. hydrophila* clearly show that the organism prefers biomass components of low molecular weight as growth substrate. The high substrate affinities of *A.*

hydrophila for arginine, ornithine and glutamate suggest that *Aeromonas* compete successfully for these compounds at low substrate concentrations.

TABLE 3 : NON-UNIFORM GROWTH RESPONSE OF *AEROMONAS HYDROPHILA* IN THE REPLICA TEST*

Compounds	Growth response of *A. hydrophila* (at 25°C for 7 days)		
	1	2	3
Amino acids			
1. Leucine	–	–	–
2. Tryptophane	+	–	–
3. Methionine	–	–	–
4. Arginine	+++	+++	+++
5. Glycine	–	–	–
6. L-Lysine	+	+	–
7. t-Ornithine	+++	+++	+++
8. t-Aspertate	+++	+++	+++
9. t-Serine	+++	+++	+++
10. Glutamate	+++	+++	+++
Carbohydrates & Alcohols			
11. D-galactose	+++	+++	+++
12. Sucrose	+++	+++	+++
13. D-Arabinose	+++	+++	+++
14. Lactose	–	+	–
15. Inositol	+	–	–
16. Maltose	+++	+++	+++
17. Fructose	+++	+++	+++
Carboxylic acids			
18. Tartric acids	+	+	+++
19. Citric acids	+++	+++	+++
20. Succinic acids	+++	+++	+
21. Solicylic acids	+	–	–
22. Oxalic acids	+	–	–
23. Pyruvic acid	–	–	+
24. Lactic acid	–	–	–
25. Salfanilic acid	–	–	–
26. Acetic acid	–	–	+
27. Formic acid	+	–	–

* Substrate were each at a concentration of 2.5 g/litre.
+++ Good growth
\+ Weak growth
– No growth

TABLE 4 : BIOCHEMICAL CHARACTERISTIC PROPERTIES OF AEROMONADS ISOLATED FROM RIVER NARMADA AT JABALPUR (AUG '93 TO MARCH '94)

	Biochemical Characters	Isolates				
		A	B	C	D	E
1.	Motility at RT	+	−	+	−	−
2.	Growth at 37ºC	+	−	+	−	−
3.	Growth at RT	+	+	+	+	+
4.	Catalase	+	+	−		
5.	Oxidase	+	+	+	+	+
6.	Huge & Leifson –Vc	−	−	−	−	−
7.	H+L Oxidative	−	−	−	−	−
8.	H+L Fert	+	+	−	−	−
9.	H+L alkaline	+	+	−	−	−
10.	Nitrate Red	−	+	+	+	−
11.	Simon's Citrate	−	−	−	−	−
12.	Urease	+	−	−	−	−
13.	Gelatin stab	+	+	+	−	+
14.	Gelatin plates	−	+	+	+	+
15.	H_2S paper	−	−	−	−	−
16.	H_2S TSI	+	+	+	+	−
17.	ONPG test	+	+	+	+	+
18.	Arginine test	+	−	+	+	+
19.	Glucose Util.	+	+	+	−	−
20.	Inositol Util.	−	+	−	−	+
21.	Lactose Util.	+	−	−	+	−
22.	Sorbitol Util.	+	−	−	+	−
23.	Sucrose Util.	−	−	+	−	+
24.	Starch Util.	+	+	+	−	−
25.	Trehalose Util.	+	+	+	−	−
26.	Methyl Red test	+	+	−	−	−
27.	Voges Proskaur test	+	−	+	−	−
28.	Indole	−	−	+	−	−
29.	Acetate	+	−	+	−	−
30.	Growth at 5ºC	+	−	+	−	−
31	Growth at 12ºC	−	−	+	−	−
	Identification Score	0.978	0.980	0.980	0.894	0.980

A = ***Aeromonas hydrophila***
B = ***Aeromonas salmonicida***
C = ***Aeromonas sobria***
D = ***Aeromonas salm***
E = ***Aeromonas caviae.***

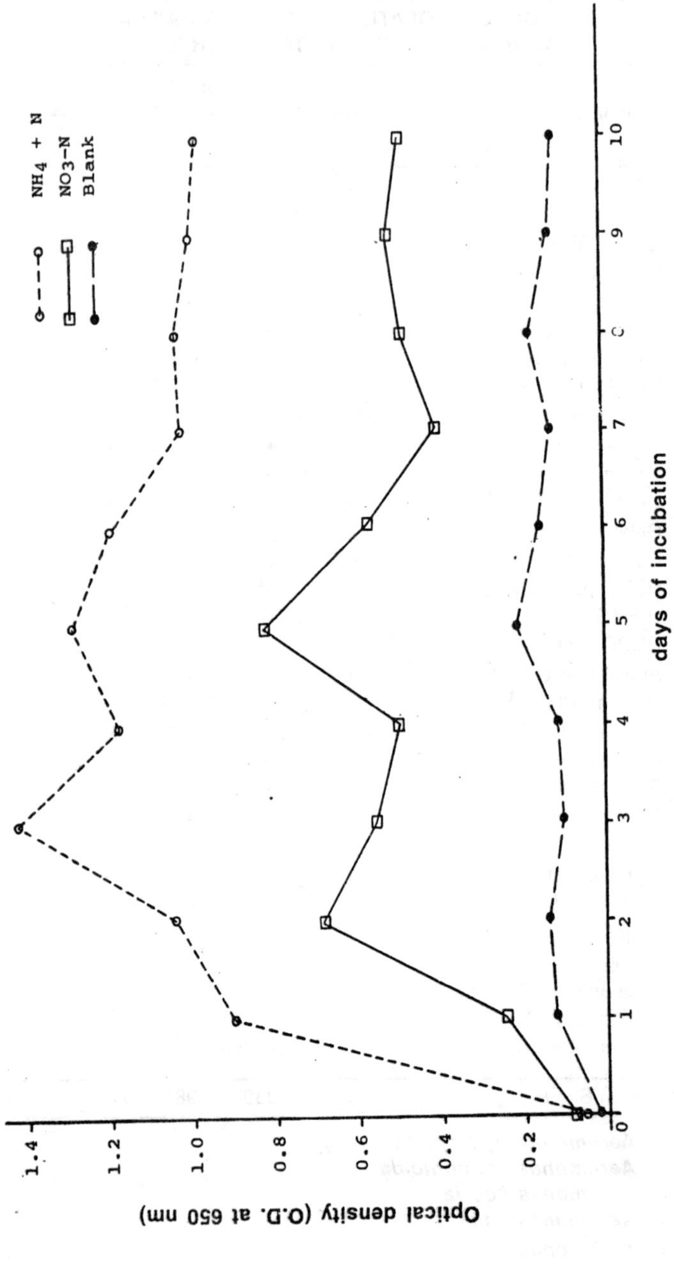

FIG 2 : EFFECT ON NH_4+N AND NO_3–N ON THE GROWTH OF *A.HYDROPHILLA* (FROM RIVER NARMADA, JABALPUR) IN A CITRATE CONTAINING MINERAL MEDIUM AT 25°C

Thus, the high incidence of clinically significant *Aeromonas* species in this aquatic region could be of public health significance for the inhabitants of this region as well as a challenge to their dependence on aquatic resources. The consideration and the data presented in this paper indicate the control of the multiplication of aeromonads in river water/ drinking water requires (i) limitation of the concentration of biomass compounds such as amino acids and long chain fatty acids in fresh waters, (ii) limitation of the concentration of other easily assimilable compounds that promote the development of bacterial biomass, and (iii) prevention of introduction of algal and bacterial biomass into the fresh water systems.

It can be concluded from the data obtained that biomass components, such as amino acids and other organic compounds can promote multiplication of aeromonads in river/ drinking water at concentration as low as a few micrograms per litre.

The finding of large number of Aeromonads especially the pathogenic forms in river water during the present investigation indicates nutrients rich condition of river Narmada, at Jabalpur. Furthermore, it can also be suggested that standards for the quality of recreational water need to be re-evaluated, as also suggested in North American Studies by Clark *et al* (1982) and Lechavallier *et al* (1982).

SUMMARY

The occurrence and distribution of *Aeromonas* species has been investigated in river/drinking waters at Jabalpur, India. Physio-chemical and bacteriological parameters were investigated. The growth of *Aeromonas* correlated with the density of total coliforms. Water temperature seems to play a significant role on their growth. Aeromonads were identified as *A. hydrophila*, *A. salmonicida*, *A. sobria*, *A. salm* and *A. caviae*. The nutritional versatility and growth response of *A. hydrophila* were also studied to determine the nature and growth promoting properties of organic compounds, which may serve as growth substrate of the organism. As an initial screening a total number of 27 different organic compounds were tested for *A. hydrophila*. Five amino acids, five carbohydrates, two carboxylic acids and one organic alcohol promoted growth of *A. hydrophila*.

The study reveals that *Aeromonas* are autochthonous members in aquatic system and are indigenous to these fresh water environments. The data obtained also indicated that biomass components can promote multiplication of aeromonads in river/ drinking water at concentration of as low as a few micrograms per litre.

ACKNOWLEDGEMENTS

The authors are thankful to CSIR, New Delhi for financial support and to Prof. S.K. Hasija, Head, Department of Bio Sciences, R.D. University for providing laboratory facilities.

REFERENCES

APHA, (1989), American Public Health Association; In standard methods for examination of water and waste water analysis (16th ed.) APHA-AWWA-WPCI, Washington D.C., 1268.

Adoni A.D. (1985), Workbook on limnology, Pratibha Publishers, Sagar, M.P. (India), p. 216.

Bryant T.N. (1989), Probabilistic Identification of Bacteria, PIB Computer Kit. Medical statistics and computing, University of Southampton, U.K. 200 p.

Buchnan R.E. and Gibbons N.E. (1984), Bergey's Manual of Determinative Bacteriology (9th ed.) The Williams & Wikkins Co. Baltimore.

Clark J.A., Burger G.A. and Sabtinos L.E. (1982), Characterization of Indicator bacteria in municipal raw water drinking water and new main water samples. Can. J. of Microbiology, **28**(6), 1002-1013.

Hajen T.C. (1979), Ecology of *Aeromonas hydrophila* in a South Carolina cooling reservoir. Microb Ecol. S : 179-195.

Kooij D.V., Visser A. and Hignen W.A.W. (1980), Growth of *Aeromonas hydrophila* at low con. of substrates added to tap water, Appl. & Env. Microbiol. **39**(6): 1198-1204.

Kooij D.V., Oranje D.J.P. and Hignen W.A.M. (1982), Growth of *Pseudomonas aeruginosa* in tap water in relation to utilization of substrates at conc. of few micrograms per litre. Appl. Environ Micro. **44**: 1080-1095.

Kooij D.V. and Hignen W.A.M. (1988), Nutritional versatility & growth kinetics of an *Aeromonas hydrophila* strain isolated from drinking water. Appl. Env. Microbiol, **54**(11), 2842-2851.

Leehavallier M.W., Evans T.M., Seidler R.J., Daily O.P., Merrel B.R., Rollins D.M. and Joseph S.W. (1982), *Aeromonas sobria* in chlorinated drinking water supplies. Microbial Ecology. **8**: 325-333 (1982).

Rippy S.R. and Cabelli V.J. (1980), Occurrence of Aeromonads in limnetic environments: relationship of the organism to trophic state. Microbiol. Ecol. **6**: 45-54.

Stanier R.Y., Palleroni N.J. and Doudoroff M. (1966), The aerobic pseudomonads a taxonomic study. J. Gen. Microbiol **43**: 159-271.

5

Development Vs Aquatic Environment: The Case of Garhwal Himalaya

Ashutosh Gautam

INTRODUCTION

The Garhwal Himalaya, comprises five districts of Uttar Pradesh namely Dehradun, Tehri, Pauri, Chamoli and Uttarkashi, lies almost centrally in the long sweep of Himalaya between 29° 26' N and 31° 28' N latitude and 77° 49' E and 80° 6' E longitude. It covers an area of about 30090 square kilometer, spreading 220 kilometer in east-west and 235 kilometer in north-south with mostly uneven topography.

The region is chiefly drained by three major river systems, the Ganga, the Yamuna and the Ram Ganga, alongwith their glacial fed tributaries as well as spring-rain fed rivers and rivulets. (Table 1 and 2)

THE UPPER GANGA RIVER SYSTEM

The Upper Ganga river system is formed by two main streams, the Bhagirathi and the Alaknanda. These two rivers meet at Deoprayag and from the confluence downstream it is called the Ganga, which flows down and emerges as the river of plains at Rishikesh. This river system drains a large area of Garhwal Himalaya except western part of Uttarkashi, Tehri and Dehradun districts. The *Bhagirathi* is an important river originating from Goumukh in Gangotri glacier and passes via thickly populated towns like Uttarkashi, Tehri and Deoprayag. It travels a long distance of 196.1 km before meeting the Alaknanda at Deoprayag. On its way it receives three main streams i.e. the Jarganga at Lanka, the Asiganga at Uttarkashi and the Bhilaganga at Tehri.

TABLE 1 : IMPORTANT RIVERS OF GARHWAL HIMALAYA

River	Origin Place	Type	Chiefly Drained Districts
Bhagirathi	Gaumukh (3900 M)	Snow fed	Uttarkashi/Tehri
Jar Ganga		Snow fed	Uttarkashi
Asiganga		Spring fed	Uttarkashi
Bhilaganga	Khatling Glacier (3590 M)	Snow fed	Uttarkashi/Tehri
Alaknanda	Alkapuri Glacier (3390 M)	Snow fed	Chamoli/Pauri
Kedar Ganga	Kedarnath Peaks	Snow fed	Chamoli
Dhauli Ganga		Snow fed	Chamoli
Rishi Ganga		Snow fed	Chamoli
Nandakini	Nanda Devi (7817 M)	Snow fed	Chamoli
Birahi		Snow fed	Chamoli
Pinder	Pindari Glacier (5500 M)	Snow fed	Chamoli
Mandakini	Dudh ki toli Range Kedarnath peak (3700 M)	Snow fed	Chamoli
Yamuna	Yamnotri (3300 M)	Snow fed	Uttarkashi/Dehradun
Tons	Northern slopes of Bunderpunch	Snow fed	Uttarkashi/Dehradun
Nayar		Spring fed	Pauri
Ram Ganga		Spring fed	Pauri
Song		Spring fed	Dehradun
Suswan		Spring fed	Dehradun
Rawsan		Spring fed	Pauri
Malan		Spring fed	Pauri
Khoh		Spring fed	Pauri

The *Alaknanda* is another important river originating from the Alkapuri glacier in Chamoli district. During its journey of 179 km it receives a large number of streams i.e. the Dhauliganga at Vishnuprayag, the Nandakini at Nandprayag, the Pinder at Karanprayag, the Mandakini at Rudraprayag and finally the Bhagirathi at Deoprayag.

This combined river as a Ganga flows down through Byasi, Gular, Lukshman Jhula, Rishikesh and Hardwar. On its way,

TABLE 2 : COURSE OF MAJOR RIVERS THROUGH GARHWAL HIMALAYA AND THEIR GRADIENT

Sl. No.	Name of the river	Upper Stretch From	Upper Stretch To	Lower Stretch From	Lower Stretch To	Distance Approx. (km)	Fall (M)	Gradient (M/Km)
1.	Alaknanda	Alkapuri Glacier (3300 M)	Vishnuprayag (1372 M)			70	1928	27.54
				Vishnuprayag (1372 M)	Deoprayag (470 M)	130	902	6.94
2.	Nandakini	Nandadevi Glacier (6000 M)	Nandprayag (914 M)	—	—	47	5086	108.21
3.	Pinder	Pindari Glacier (5500 M)	Karanprayag (795 M)	—	—	74	4705	63.58
4.	Mandakini	Kedarnath (3700 M)	Rudraprayag (610 M)	—	—	95	3090	32.53
5.	Bhagirathi	Goumukh (3900 M)	Uttarkashi (1158 M)			120	1654	13.78
				Uttarkashi (1158 M)	Deoprayag (470 M)	145	686	4.73
6.	Bhilangana	Khatling Glacier (3590 M)	Tehri (630 M)	—	—	135	2960	21.93
7.	Yamuna	Yamnotri (3300 M)	Dakpathar	—	—	150	2600	17.33
8.	Ganga	—	—	Deoprayag (470 M)	Hardwar (294 M)	80	175	2.19
9.	Nayar	—	—	Satpuli (680 M)	Byasghat (415 M)	30	265	8.83

Singh *et al* 1993

further it receives eastern and western Nayar, Gular, Song, Suswan etc.

THE YAMUNA RIVER SYSTEM

The Yamuna has its origin from the Yamunotri glacier. It travels a long distance of 150 km before reaching Pounta Sahib and receives a large number of streams on both the sides when passing across Purola, Barkot, Kalsi, Lakhwar, and Dakpathar. The Tons, a major tributary of Yamuna originates from the northern slopes of Bunderpunch, joins it below Kalasi.

THE RAM GANGA RIVER SYSTEM

Basically it is a river system of Kumaon region. In Garhwal, it is draining only south-western part of Pauri district.

WATER QUALITY OF AQUATIC ENVIRONMENT

The rivers of Garhwal Himalaya are of prime importance for the people of India. The water quality of the rivers varies from time to time and place to place, due to the interaction of local factors. Apart from these the human interaction is most effective.

In Garhwal, ablution at major pilgrimage centers, tourist spots and various other activities of our society along the upper stretch of the Ganga are responsible for the changing complexion of rivers in Garhwal. Disposal of dead bodies, animal carcasses, untreated sewage are the source of organic and biological pollution, while road construction, dam construction and deforestation etc. are the source of sediment pollution in river water. Rapid population growth and tourism development have contributed in over-exploitation of the limited resources leading to escalated water pollution (Gautam 1988). Although the water of Garhwal's river is blessed with self curative power, high rate of pollutant influx is resulting in the changing complexion of rivers. It is apparent from the behaviour study of various characteristics of river water that the holy riverine system of Garhwal now undergoes the process of deterioration (Gautam and Singh 1995).

MAJOR DEVELOPMENTAL ACTIVITIES

Alongwith the booming of tourist industry, the urbanization and human population have grown manifold during the last few years. As a result, large amount of sewage and other organic wastes are carried into the rivers every year; excessive erosion and silting have increased the sedimentation rate in the rivers, this in turn clogs the gills of fishes and remove their food from the river bed, causing depletion in fish population. Fish population of the river Bhagirathi was greatly reduced during the year 1978 when a blockade formation took place across the river course near Kathuli Gad (Sharma and Singh 1980). Major development related activities, conducted by human population to fulfil their requirements along the river courses in Garhwal Himalaya, responsible for the deterioration in aquatic environment are as under:

Construction

The construction is a basic requirement for any development, but the construction of roads, dams and buildings in this region is unscientific and unethical. Blasting of rock structures and dumping of debris/cutsoil leads to a higher amount of solids in the river water, causing immedicable damage to the lotic aquatic ecosystem.

Deforestation

Large scale exploitation of forest resources is done to fulfil the ever-increasing demand of agricultural land, fuel, raw material and infrastructure facilities. Most of the region is already prone to land slides due to geodynamic unstability and reduction in vegetal cover is further aggravating the situation by increased soil erosion.

Modern Agriculture

Pesticides and fertilizers both certainly helps to increase agriculture production, but their increased quantities and consistent uses can have serious ecological consequences on the aquatic environment. These find their way to the nearby stream through surface runoff from the agriculture fields. The major effects are the fish kill and eutrophication.

Stone Collection

Due to the ease availability, stone is the most economic building material for the people of this region. It is also a basic source of economy, as many stone crushing units are under operation in Garhwal and its surroundings. Collection of stones from the river bed is growing day by day with the increasing demand of the building material. Now it is becoming an enigma which is growing rapidly.

Sewage

Dumping of sewage without any treatment is a common practice in Garhwal, deteriorating water quality of the rivers slowly but surely (Nautiyal and Rachna 1992, Gautam *et al* 1989 and 1993). The increasing input of sewage can cause a greater damage to the aquatic environment of Garhwal, therefore, there is an urgent need to adopt suitable treatment technology for controlling sewage pollution.

PRINCIPAL POLLUTANTS

Any substance which on addition to the environment changes its natural characteristics is called environmental deteriorator or pollutant. In a simple term pollutant is that material which causes pollution in the environment. According to the Environmental Pollution Panel of the US President's Advisory Committee, "The environmental pollution is the unfavourable alteration of our surroundings wholly or largely as a byproduct of mans' actions through direct or indirect effects of changes in energy patterns, radiation levels, chemical and physical constitution and abundance of organisms.

US Environmental Protection Agency classified aquatic pollutants into eight categories, as a part of the development various pollutants which are responsible for the deterioration in the aquatic environment of Garhwal Himalaya differ slightly. These can be classified into six categories i.e. (i) Oxygen demanding wastes; (ii) Disease causing agents; (iii) Sediments; (iv) Nutrients and Salts; (v) Metals and (vi) Others like pesticides, herbicides, insecticides, fungicides, detergents and natural toxins. The details are given in Table-3 (Gautam, 1990).

Considering all six categories of pollutants, pollution in the rivers of Garhwal may be characterised by three broad groups i.e. Physical deterioration; Chemical deterioration and Microbiological deterioration. Further, chemical deterioration can be sub classified into two sub groups i.e. Organic deterioration and Inorganic deterioration (Gautam and Singh 1995).

TABLE 3 : PRINCIPAL POLLUTANTS

Activity and Direct Cause	Principal Type of Pollutants	Environmental Impacts	Magnitude
URBANIZATION			
(Population Growth)	Untreated Domestic Sewage	High Bacterial Content, Virus, and Parasites. Increased BOD, Higher Nutrient Concn.	High in Both Main Stream and Tributaries
(Construction)	Sediments	Increased Sedimentation Rate & Turbidity	– Do –
(Domestic Cleaning)	Detergents	Direct Toxicity	Low in Main Streams
(Public Health Measures)	Chemicals e.g. D.D.T. & residues	Direct Toxicity in Receiving Waters	Low in Main Streams
AGRICULTURE			
(Agricultural Chemicals)	Residues of Pesticides and Fertilizers	Increased Toxicity, B.O.D. and Nutrient Level	High in Tributaries
(Soil Erosion)	Sediments	Increased Sedimentation Rate, Turbidity	High in Tributaries
DEVELOPMENT			
(Road, Dam and Building Construction)	Dumping of Sediments	Increased Sedimentation Rate, Turbidity	Very High in Both
(Deforestation) High Rate of Land Slides, Slips and Soil Erosion	Sediments	Increased Sedimentation Rate and Turbidity	Very High Both Main Stream and Tributaries
(Collection of Stones)	Sediments, Rising Temperature, Shallowness	Direct Effect on the Biota and Physico-Chemical Characters	Low But Growing Rapidly
(Quarrying) Increased Soil Erosion)	Sediments	Increased Sedimentation Rate, High Turbidity	Low

Sources : Impact and Magnitude

IMPACT ASSESSMENT

A slight change in the surrounding activity and or environment can cause a greater harm to the water quality. The construction activity, deforestation and blasting of rock structures is a potential cause behind the physical deterioration in terms of turbidity, total solids and transparency. The ingression of silt and other solid wastes from the construction can exert a profound effect on the temperature, free CO_2, and dissolved oxygen as experienced by Singh, 1987. Silt, the major source of physical pollution in the Garhwal Himalaya, hinders the penetration of sunlight. Which inturn reduces photosynthetic activity and phytoplanktonic growth (Nautiyal, 1984 and 1985). Fishes, the major victims, get affected due to non-availability of food material as well as due to blocking of the gills.

Sewage, one of the most prominent sources of environmental deterioration in Garhwal Himalaya, have serious ecological consquences. As experienced, it enhanced the concentration of organic material, nutrients and salts.

The construction of dam across the lotic systems adversely affect water current, water temperature, dissolved oxygen, free CO_2, etc. According to Singh 1987 and Nautiyal and Singh 1989 the obstruction due to the dam construction across the river course have adversely affected the free movement of fishes to spawn and other related life sustaining activities in their study zone. The other most important aspect, may alter the existing flora and fauna of the lotic environments with high water velocity/current and low temperature, is the stagnation of water in the form of reservoirs. Therefore, the availability of food material, types of food material and feeding stimuli-feeding receptors will certainly not in the coherence with the existing fish population, particularly the situation is highly sensitive for strictly bottom feeders and dwellers (Singh *et al.*, 1993).

Pesticides which are used in modern agriculture can cause short term as well as long term effects like-instant killing of the living organisms and biotransformation, because these are not easily biodegradable and can sustain for a longer period. On the other hand ingress of the fertilizers can cause adverse changes in the community structure.

SUGGESTIONS

The foregoing account states that the people of this region are using rivers as a natural dustbin for dumping liquid as well as solid wastes without taking care of hazards. To take a step towards sustainable development there is a dire need to abandon all undesirable activities of the community. With the below stated suggestions we can maintain our aquatic environment more productive without ecological damage.

- Preparation of authentic data base;
- Continuous monitoring of environmental parameters;
- Impact assessment study for various developmental projects;
- Proper sewage treatment before disposal;
- Afforestation along the catchment and on the threatened slopes;
- Peoples' participation; and
- Environmental education.

ACKNOWLEDGEMENTS

Thanks are due to Professor Shashikant, University of Jammu and Dr. S.R. Mishra, Bhind (MP) for providing an opportunity to express our great concern over the aquatic environment of Garhwal Himalaya in the form of a paper. Thanks are also due to Department of Environment for providing financial assistance. Thanks are due to Dr. Krishna Gopal, IIRC, Lucknow and Dr. P.K. Goel, Y.C. College of Science, Karad for valuable interaction and support.

REFERENCES

Gautam Ashutosh (1988), Protect eco-system of Garhwal Himalaya, Yojana **32**(10): 28 & 40.

Gautam Ashutosh (1990), Ecology and Pollution of Mountain Waters. Ashish Publishing House, New Delhi, India.

Gautam Ashutosh (1992), Some consideration on stone removal from river basin and resultant effects. J. Ecobiol **4**(1): 55-58.

Gautam Ashutosh, Joshi V.P. and Sati O.P. (1993), Physico-chemical characteristics of sewage and its impact on water quality of Alaknanda at Srinagar (Garhwal). J. Ecotox. Env. Mon. **3**(1): 61-63.

Gautam Ashutosh and Singh N. (1995), Water quality of lotic aquatic environment in Garhwal Himalaya Towards deterioration : A review. *In*: Recent Researches in Aquatic Environment (Gautam Ashutosh and Agarwal N.K. ed.) pp. 67-76, Daya Publishing House, New Delhi.

Gautam Ashutosh, Singh N. and Juyal C.P. (1989), On chemical parameters and role of sewage as pollutant in river Bhagirathi at Uttarkashi (Garhwal Himalaya) Him. J. Environ. Zool. **3**: 262-264.

Nautiyal P. (1984), Studies on the riverine ecology of torrential water in Indian uplands of Garhwal region. II. Seasonal fluctuation in diatom density. Proc. Indian Acad. Sci. (Animal Sciences) **93**: 671-674.

Nautiyal P. (1985), Studies on the riverine ecology of torrential water in Indian uplands of Garhwal region. I. Seasonal variations in percentage occurrence of planktonic algae. Uttar Pradesh Journal of Zoology **5**: 14-19.

Nautiyal P. and Rachna (1992), Civic effluents, impact assessment and water quality perspectives in mountain river of Garhwal Himalaya *In*: Aquatic Environment (Gautam Ashutosh ed.) pp. 22-35. Ashish Publishing House, New Delhi, India.

Nautiyal P. and Singh H.R. (1989), River projects and the endangered hill stream Mahseer *Tor Putitora* in the Indian uplands of Garhwal Himalaya, a critical evaluation. Him. J. Environ. Zool. **3**: 129-134.

Sharma R.C. and Singh H.R. (1980), Impact of Bhagirathi blockade on its river ecosystem. Johsard **4**: 35-38.

Singh H.R. (1987), Environmental problems in Garhwal Hills *In*: Science, Development and Environment (Agarwal V.P. ed.): 37-45 Society of Biosciences, Muzaffarnagar, India.

Singh N., Bahuguna S.N. and Bhatt K.C. (1993), The profile of river ecosystem, food and feeding habits of hill-stream fishes and consequences of recent environmental degradation in Garhwal Himalaya. Acta Ichthyologica et Piscatoria XXIII (1): 3-30.

6

Hydrological Features and Diatom Flora of Pulicat Lake

R. Ramanibai & D. Sasikala

INTRODUCTION

Diatoms are the nature's gift to the paleontologists to ascertain historical background of types of aquatic medium and their time of existence. Innumerable contributions were made and are being added in the field of diatom research and its related areas. The abundance and distribution of diatom taxa were studied in a brackish water lagoon 'Pulicat lake' in South India. Southern part of the lake comprise more number of diatoms than the central and northern part. The availability of diatom in this area may be due to its interlink with the bay, canal and rivers. Similarly monsoon also alter the density and distribution of diatom flora in the lake where more number of individuals were present during summer month.

Diatoms are highly sensitive to water proporties and their existence is mainly influenced by three factors the pH, salinity and temperature (Schrader *et. al.*, 1993). Generally, combinations of different environmental parameters are responsible for the nature and distribution of flora and fauna in a waterbody. The sensitivity of diatoms for changes in the water quality has prompted paleoliminologists to reconstruct past environments. They rely on the structure of diatom community deduced from examining the materials recovered from core sediments of lake or coastal environments. Such studies are interesting and might reveal new information especially from coastal environment.

The taxonomy of lake water diatoms have not received much attention as those from the seas around our country. The pulicat lake is a large coastal lagoon with many unique features on the east coromandal coast. The diatoms from this lake is yet to be investigated in detail. A study of diatom community from the core, surface sediments and water column may be interesting and also provide new accounts of lentic diatom species from this part of India.

MATERIALS AND METHODS

The Pulicat lake is the second largest coastal (brackish water) lagoon, lies in between Tamil Nadu and Andra Pradesh and has also connection with the Bay of Bengal. Water and plankton samples were collected in the pulicat lake covering a distance of 85 km. The lake has been divided into four parts namely southern, northern, central and channel part and sampling stations were fixed in these parts for water and diatom collections. (Fig. 1). Biannual sampling pattern was adopted for sampling, one during post monsoon and another during summer months. Diatom samples were collected by using plankton net of 0.0625 mm mesh size. Subsurface and sediment core samples of about 3 m depth were also collected from different locations. The samples were immediately transferred to the laboratory for further analysis.

The isolation of diatoms from the subsurface and core sediment samples were dried in shade. The samples were first oxidised with HCl and HNO_3 (1:1) mixture to remove the organic substances (Gray, 1965). Subsequent treatment with heavy liquid (Cd I_2 and KI) was done to isolate the diatoms from other planktonic debris. This method was adopted for water samples also. Before mounting, a drop of pellet was put on a cover-slip and kept in the oven for drying and mounted in DPX. The diatoms were identified and enumerated using light microscope (10 x 10 and 10 x 40 x magnification) with the help of standard keys. Water samples were analysed for pH, EC and salinity in the lake site using portable field kit. The standard methods (APHA) were adopted for the analysis of major anions and cations.

RESULTS

Water Chemistry

The results of analysis of water samples during post monsoon (Feb. 94) and summer (June 94) are given in Table 1

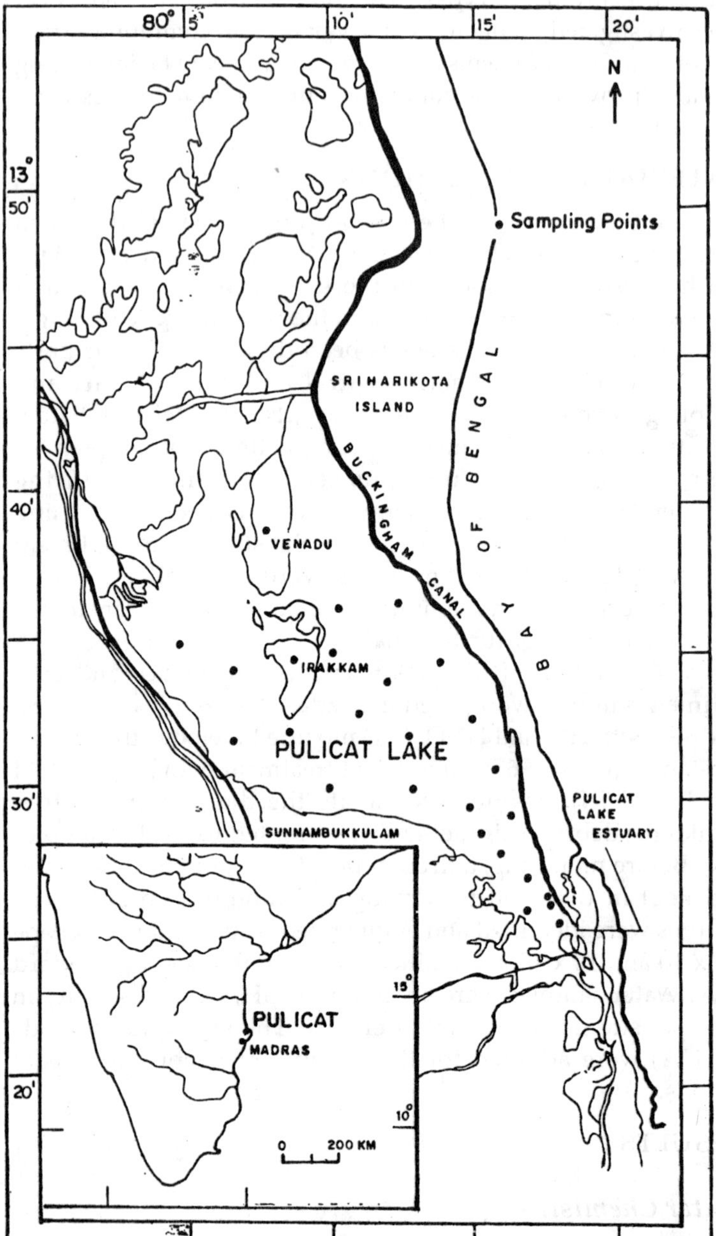
80°
5'
10'
15'
20'
13°
50'
40'
30'
20'
N
Sampling Points
SRIHARIKOTA ISLAND
BUCKINGHAM CANAL
BAY OF BENGAL
VENADU
IRAKKAM
PULICAT LAKE
PULICAT LAKE ESTUARY
SUNNAMBUKKULAM
PULICAT
MADRAS
15°
10°
0 200 KM

FIG. 1

as average for the four prominent zones of the lake - the Northern part, Southern part, Central part and channel part (Fig. 1). There are significant climatic changes during these periods which also got reflected in the water chemistry. Water temperature did not show wide variations between seasons but pH showed significant changes. pH was generally above 8.0 however, the Southern and Northern part showed a decline during June 94 due to the influence of fresh water supply. The EC and salinity were generally higher on the Northern and central part than the other regions.

The changes in chloride concentration ranging from 19632 to 26913 ppm appear complex as it is dynamically affected by three processes, the tidal influence, fresh water inflow and the evapoconcentration and no major cause could be attributed in any part of the lake on its variation. The sodium concentration varying from 9385 to 13281 ppm also showed a similar behaviour (Table 1). The sulphate concentration was high in the southern part of the lake where as many ions showed higher concentration in the central part.

The carbonates and bicarbonates were low in Pulicat lake waters than other elements. The bicarbonate showed a wide variation during the post monsoon season from 33 to 173 ppm at various part of the lake. In some stations, the concentration was more than the sea water suggesting evaporation. The total hardness of the Pulicat lake waters were also high varying from 784 to 2072 ppm. It is especially high in the central part during summer period. The sediments of Pulicat lake at many parts are rich in gypsum which may be the cause for higher calcium content of the water (Krishnaswamy, 1979).

DIATOMS IN SURFACE WATERS

A total number of 41 common diatom species belonging to 24 genera were enumerated from water, sediment and core samples of the Pulicat lake. The list of the species recorded are given in Table 2. Diatoms appeared more abundant in the surface waters than in surface sediments (Table 2) and very low in the core sediments.

It appears that the occurrence of diatoms are related to temperature fluctuations. In central part the temperature was at its maximum (34°C) which also recorded more number of

TABLE 1 : PHYSICO-CHEMICAL CHARACTERISTICS OF PULICAT LAKE WATER DURING POST MONSOON AND SUMMER PERIOD 1994

Region	Season	Temp °C	pH	EC mS/Cm	Salinity %	Alkalinity	Total Hardness	Na	K	SO_4	Cl
						ppm					
Northern Part	Feb	30.9	8.25	16.05	36.7	163	3799	12476	1141	1514	20993
	June	28.8	7.95	16.54	22.6	322	9110	2939	508	23962	25771
Central Part	Feb	30.1	8.26	14.11	35.2	109	3331	9888	870	1428	18934
	June	29.0	8.00	16.80	21.8	316	9393	2593	481	22162	25161
Southern Part	Feb	30.6	8.31	14.40	35.9	154	3799	11108	911	1558	21584
	June	29.0	7.80	14.37	23.1	316	8792	2757	532	24374	25745
Channel Part	Feb	29.0	8.25	14.39	32.5	147	3805	10984	1003	1495	19508
	June	29.3	8.00	14.70	22.3	299	7373	2602	452	23998	24839

diatoms. Similarly at low pH value the percentage occurrence of diatom species was very high. The alkaline nature of the lake water exhibit more number of diatom fauna.

During post monsoon period the abundance of diatoms was more when compared to June collection. The most dominant forms of diatom sampled during post monsoon are ***Cocconeis heteroidea,*** and ***Coconeis*** sp. The dominant forms found in early summer are ***Cocconeis heteroidea, Cocconeis*** sp, ***Grammatophora marina, Grammatophora oceanica*** and ***Amphora*** sp. The moderate forms are ***Amphora*** sp, ***Coscinodicscus*** sp, ***Diploneis*** sp, ***Entopyla australis, Mastogloia*** sp, ***Navicula*** sp, ***Grammatophora Oceanica,*** and ***Auliscus sculptus.*** The less frequent are ***Grammatophora marina, Cocconeis heteriodea, Coscinodiscus asteromphalus, Auliscus sculptus, Plagiogramma interruptum, Rhopalodia gibberula, Auliscus*** sp, ***Conscinodiscus*** sp, ***Diplonels*** sp, ***Plagiogramma staurophorum, Amphora*** sp, ***Grammatophora Oceanica, Deploneis Wessflogii, Mastogloia*** sp, ***Plagiogramma obesum, Plagiogramma*** sp and ***Triceratium*** sp. (Table 2).

During early summer period, the density of diatom species and the dominant forms are more when temperature is high in channel part (between 30-35° C). The common diatoms found in these three stations are ***Grammatophora marina*** and ***Cocconeis*** sp. The density of diatom is very less and no dominant species are found when temperature is low. During this early summer period, the dominance and the density of diatom is maximum at a lower pH of 8.13. The most dominant forms found here are ***Cocconeis heteroidea, Cocconeis*** sp, ***Grammatophora marina*** and ***Grammatophora oceanica.***

In the summer period, the density of diatom is less and the number of genus and the number of species were more in almost all the stations. The density of diatom and the number of genus and species are higher at channel part whereas the density and the number of genus and species are less at Northern part. The common diatoms found dominant are ***Actinocyclus octonarius, Amphora*** sp, ***Grammatophora Oceanica*** and, ***Mastogloia*** sp. The moderate forms are, ***Coscinodiscus*** sp, ***Cocconeis heteroidea, Diploneis*** sp, ***Paralia sulcata*** and ***Plagiogramma*** sp, ***Aulacoseira distans, Navicula*** sp, ***Licmophora*** sp, ***Aulicus*** sp, ***Buddulphia aurita, Amphora turgida, Amphora spectabilis, Actinocyclus***

kuetzinglii. There is no dominant form here. In all other parts of the lake only a least number of forms are found. Some of the common forms are, ***Amphora crassa, Coscinodiscus asteromphalus, Mastogloia*** sp, ***Navicula*** sp, ***Cocconeis*** sp, ***Cocconeis heteroidea, Disploneis*** sp and ***Grammatophora marina.***

The samples collected during summer period recorded high diatom density, more number of genus and species and its abundance is also more at channel part. The pH is not showing much variation comparatively and the diatoms found are more at pH 8.0 and low at pH 7.5 where only the diatom ***Grammatophora marina*** is the most dominant form.

DIATOMS IN SEDIMENTS

The density of diatom and the dominant forms are more in summer months when compared with the post monsoon periods but the number of genus and species are more or less the same (Table 2).

In summer, the density of the diatom is higher at Northern part of the lake. In all other parts the density is more or less moderate. The number of abundant forms is also more during the summer periods when compared to the early summer periods. The dominant forms found in Northern part are, ***Cymatosira lorenziana, Diploneis*** sp., ***Grammatophora Oceanica, Cocconeis*** sp., and ***Navicula*** sp., are the most abundant forms. The moderate forms found here are ***Rhopalodia*** sp., and the diatom ***Diploneis weissfloggii*** is found less. The diatoms ***Diploneis*** sp., and ***Grammatophora Oceanica,*** are found in moderate amount. The ***Auliscus Sculptus, Plagiogramma*** sp., and ***Triceratium*** sp., are found in less number (Table 2).

In all the other stations, the dominant forms are ***Amphora*** sp., ***Diploneis*** sp., ***Grammatophora oceanica, Amphora turgida, Auliscus sculptus, Coscinosdiscus*** sp., ***Navicula*** sp., ***Grammatophora marina, Mastogloia*** sp., ***Plagiogramma obesum, Paralia sulceta, Rhopalodia*** sp., ***Cocconeis*** sp., and ***Rhadbonema punctatum. Grammatophora oceanica*** is the most abundant form (Table 2).

TABLE 2 : LIST OF COMMON DIATOM FLORA PRESENT IN THE WATER, SURFACE AND CORE SEDIMENT OF THE PULICAT LAKE DURING 1994

Sl. No.	Diatom	February			June		
		W	S	C	W	S	C
1.	Actinocyclus kuetzingii	–	1	–	1	–	–
2.	Actinocyclus octonarius	–	1	–	15	–	–
3.	Amphora crassa	–	–	–	1	–	–
4.	Amphora turgida	–	2	–	2	15	–
5.	Amphora spectabilis	–	1	–	1	–	1
6.	Amphora sp.	12	2	1	12	55	–
7.	Aulacoseira distans.	–	4	–	1	4	–
8.	Auliscus sculptus	3	3	–	–	33	–
9.	Auliscus spp.	1	1	–	1	–	–
10.	Biddulphia aurita	–	2	–	1	–	–
11.	Climacosphaenia moniligera	–	1	–	1	–	–
12.	Cocconeis heteroidea	56	–	–	4	1	–
13.	Cocconeis sp.	129	12	–	–	67	–
14.	Coscinodiscus asteromphalus	1	1	–	–	–	–
15.	Coscinodiscus excentricus	–	1	–	1	–	–
16.	Coscinodiscus sp.	4	3	–	4	7	–
17.	Diploneis weissflogii	1	–	–	–	8	–
18.	Diploneis spp.	4	–	–	6	69	–
19.	Entopyla australis	3	1	–	–	–	–
20.	Fragillaria cylindrus	–	–	–	3	1	–
21.	Frickea lewisiana	–	–	–	2	1	–
22.	Grammatophora marina	63	6	–	11	32	–
23.	Grammatophora oceanica	22	55	–	5	166	–
24.	Licmophora ehrenbergii	–	1	–	1	–	–
25.	Licmophora sp.	–	2	–	1	–	–
26.	Mastogloia sp.	5	3	–	25	41	–
27.	Navicula algida	5	1	–	–	–	–
28.	Navicula retusa	–	3	–	–	3	–
29.	Navicula zostereti	–	1	–	1	–	–
30.	Navicula sp.	9	10	–	15	69	–
31.	Paralia sulcata	–	–	–	2	9	–
32.	Plagiogramma interruptum	1	58	–	–	–	–
33.	Plagiogramma obesum	1	2	–	–	22	–
34.	Plagiogramma staurophorum	2	52	–	–	–	–
35.	Plagiogramma sp.	1	1	–	3	4	–
36.	Plagiotropis lepidoptera	–	1	–	1	–	–
37.	Rhopalodia gibberula	1	1	–	2	3	–
38.	Rhopalodia sp.	–	1	–	1	16	–
39.	Rhadbonema punctatum	–	9	–	1	23	–
40.	Triceratium sp.	4	3	–	–	5	–

W : Water Sample
S : Surface Sediments
C : Core Sediments

DIATOMS IN CORE

The diatoms from core samples are highly limited. Diatoms could be identified only at the top portion of the core samples. The diversity and density of the diatoms were also less when compared with the surface sediments and water column. Only four forms are found to occur frequently ; ***Grammatophora oceanica, Navicula lyra, Navicula brasiliensis, Pseudotriceratium punctatum*** (Table 2).

DISCUSSION

There appears to be a significant difference between the water quality at different parts of the Pulicat lake. The climate as well as the influence of sea and fresh water supply appear to influence the water quality coupled with the evaporation and isolation of water masses simultaneously. These are significant aspect that may influence the density and diversity of diatom flora in Pulicat lake. Therefore a further analysis of water chemistry data was done. An attempt was made to classify the sampling stations in terms of water quality during February '94 and June '94.

The cluster analysis was done using the sampling stations as observations and different water quality parameters as variables. Separate analyses were done for the February (post monsoon) and June '94 (Summer) samples. SPPS PC+ a statistical package was used to perform these analysis in a Personal Computer.

The results are shown in Fig. 2 as a dendrogram. The appearance of the dendrogram is quite different between the two seasons. In February '94 three groups of stations can be seen. The stations falling near the mouth of the lake except station I and 2, the stations on the northern part and channel part of the lake falling under fairly separate groups with few overlaps. The analysis since based on the water quality parameters suggest a suitable difference in quality between these locations. The channel part is near the sea mouth and during February the bar mouth is wide and open and therefore the tides can be seen influencing the water quality. The northern part is under the influence of the rivers draining into the lake with fresh water supply and therefore appear close to freshwater conditions. The

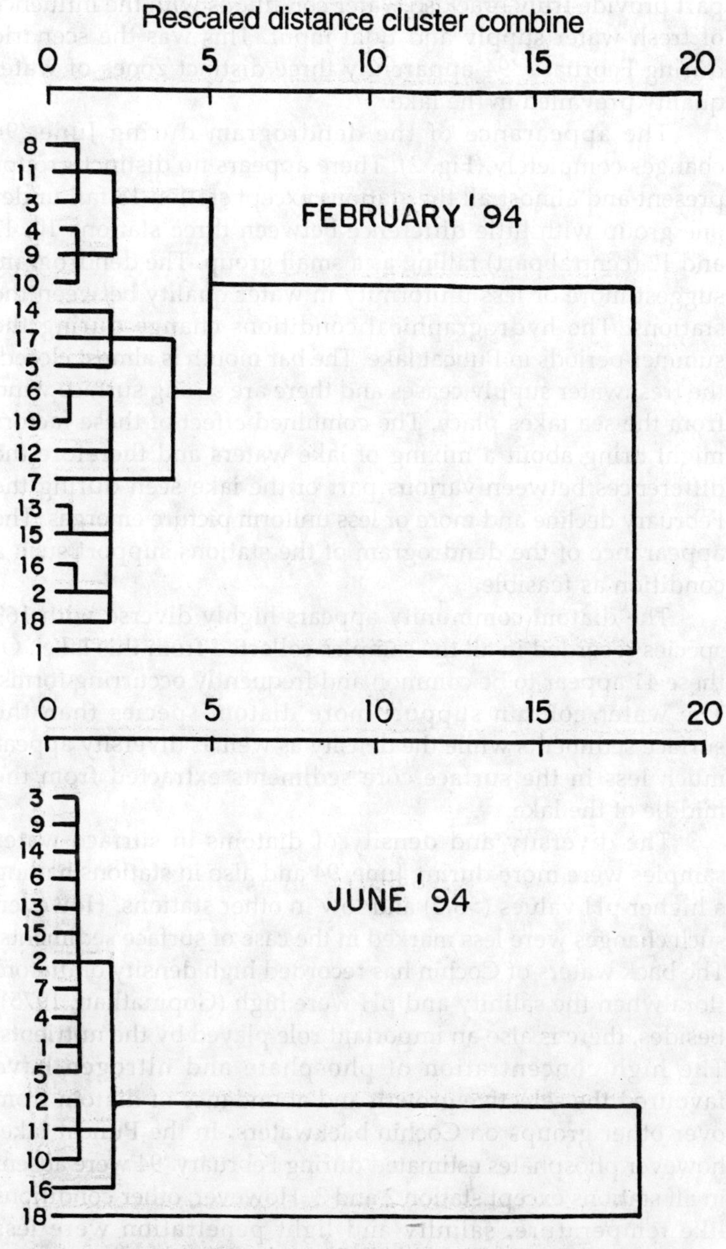

FIG. 2 : DENDROGRAM OF SAMPLING LOCATIONS IN PULICAT LAKE GENERATED USING WATER QUALITY DATA

salinity was the lowest in this part of the lake. The southern part provide truly brackish water conditions with the influence of fresh water supply and tidal input. This was the scenario during February '94 apparently three distinct zones of water quality prevailed in the lake.

The appearance of the dendrogram during June '94 changes completely (Fig. 2). There appears no distinct groups present and almost all the stations except station 18 fall under one group with little difference between three stations 10, 11 and 12 (central part) falling as a small group. The dendrogram suggest more or less uniformity in water quality between the stations. The hydrographical conditions change during the summer periods in Pulicat lake. The bar mouth is almost closed, the freshwater supply ceases and there are strong surface wind from the sea takes place. The combined effect of these factors might bring about a mixing of lake waters and therefore the differences between various part of the lake seen during the February decline and more or less uniform picture emerges. The appearance of the dendrogram of the stations support such a condition as feasible.

The diatom community appears highly diverse with 169 species recorded in all the samples collected from this lake . Of these 41 appear to be common and frequently occurring forms. The water column support more diatom species than the surface sediments while the density as well as diversity appear much less in the surface core sediments extracted from the middle of the lake.

The diversity and density of diatoms in surface water samples were more during June '94 and also in stations having a higher pH values (>8.0) and low in other stations. However, such changes were less marked in the case of surface sediments. The back waters of Cochin has recorded high density of diatom flora when the salinity and pH were high (Gopinathan, 1975). Besides, there is also an important role played by the nutrients. The high concentration of phosphate and nitrogen have favoured the selective growth and abundance of diatom from over other groups on Cochin backwaters. In the Pulicat lake, however phosphates estimated during February '94 were absent in all stations except station 2 and 3. However, other conditions like temperature, salinity and light penetration were less favourable in Pulicat lake during February '94. Therefore even when the nutrients were available higher growth rates could not have been obtained and may be the reason for low diatom density noticed in February '94. The vigorous diatom growth during summer may have depleted phosphate in many stations noticed in Pulicat lake during June '94.

In the river Ter also water salinity and pollution due to urban has been identified as two main factors controlling the distribution and abundance of diatoms as a tools. When using PCA and Canonical Correlation Analysis Water pollution appear to favour diatom assemblages with dominance of ***Nitzchia*** and ***Gromphonema genus.*** Whereas the water salinity, chlorides, sulphates, nitrates and silicates identified by a principal component axis has significant correlation with the genus ***Navicula*** in river Ter (Sabater and Tomas, 1990). Though species level analysis has not been done in this study the relationship between physical and chemical parameters and diatom density in Pulicat lake agrees with these findings.

In co-ordination of 40 lakes on eastern Finland using diatom assemblages of surface sediments, five groups of lakes are defined that have definite characteristics in terms of artificial drainage, diffuse nutrient loading and acidification (Havttunen and Merilainon, 1983). The role of nutrients as in the case of Pulicant appear to be the cause of shifts from one group to the other group of lakes.

The upwelling of nutrients during March may be under the influence of the South west monsoon takes place in the Bay of Bengal has been attributed to the bloom of diatoms noticed in Gopalpur coast (Choudhury and Panigraphy, 1989). The depletion of silicates by a ten fold decrease has been suggested as possible cases for the termination of bloom. However, in this study the silicates have not been estimated whose dynamics may also have affected the diversity and density of diatom flora in Pulicat lake.

The gradients in water quality is not readily discernible in Pulicat lake except the temporal differences recorded in February and June '94. Such a gradient existing in the Mississippi salt marsh and its influence on the structural differentiation of diatom communities has been explained by using Canonial Correlation Analysis (Sullivan, 1982).

In the Pulicat lake also the changes in water quality as evidenced by the dendrogram of sampling stations influenced the density and diversity of diatom flora. The influence of the monsoon and flow of fresh waters into the lake generated three distinct zones of water quality. However, this resulted in lowering of pH as well as salinity in many parts of the lake and were not conducive to diatom growth. With the advancement of summer, the water quality zones becomes less distinct and the marine influence increased in extent. This also raised the pH of the waters and the salinity triggering the growth of diatoms.

SUMMARY

The habitat preferences, abundance and distribution of the diatom flora were studied in a brackish water lagoon Pulicat lake adjoining the Bay of Bengal. Surface water samples at different locations were collected and analysed for water quality parameters. The analysis suggest the occurrence of marine - brackish - fresh water zones within the lake. Chloride and sulphate level seem to be high enough (>20,000 mg/l) to enable the occurrence of many forms of diatoms. The complete absence of phosphate in many locations of the lake reveal the high productivity of the lake. Euryhaline brackish water and euryhaline marine forms of centric and pennate diatoms are more in number in the samples collected during June '94. Seasonal variations appear to have significant effect on percentage of diatom assemblages. Common genera for both water and sediment samples comprise ***Nitschia*** sp., ***Grammatophora*** sp., ***Coscinodiscus*** sp., ***Navicula*** sp., ***Amphora*** sp. and ***Biddulphia*** sp. Observations and analysis done so far indicate that the occurrence of diatoms is not uniform among the habitats surveyed and have regional as well as seasonal specificity. In addition, various physical and chemical properties influence the spatial and temporal distribution of diatoms in Pulicat lake.

REFERENCES

APHA (1981), Standard Methods for the examination of water and waste water, 16 Ed, American Public Health Association, Washington, D.C.

APHA (1986), Standard Methods for the examination of water and waste water, 16th edition, American Public Health Association, Washington, D.C.

Choudhury, S.B. & R.C. Panigraphy, Ind. Jour. of Mar. Sci. **18** (1989); 204-206.

Gopinathan, C.P. Bull. Dept., Mar. Sci. Univ. Cochin, VII, **4** (1975), 995-1004.

Havttunen, P. & Kouko Merilainen. Hydrobiologia **109** (1983), 91-97.

Jane Gray, Extraction Techniques. Edited by David Raup (1965).

Krishnaswamy, S. 1979. India's Mineral resources: Oxford & IBH Publishing Co., 658 p.

Sabater, S., X. Tomas. Nova Hedwigia 51 (1990), 165-166.

Schrader, H. Inger. Lindstrom. Swanberg. *Hydrobiologia* 269/270 (1993), 129-135.

Sullivan, M.J. Jour. Phycol. **18** (1982), 130-133.

7
Biodiversities and Reclamation of Baraila Lake : A Wetland Study

Awadhesh Kumar Mishra

INTRODUCTION

The study of lakes has developed into a popular hydrological pursuit in recent years, and there are those who would encourage even greater expansion in this area. Perhaps first, however, we would be well advised to pause and reflect on some of the more timely developments of current limnological research. One of these is in the area of biodiversities and reclamation of Baraila lake (a wetland study).

The term 'wetland' deals with the diverse ecosystems like swamp, marsh, bogs, lakes, and woodland etc. The position of a wetland is inbetween terrestrial and deepwater aquatic systems.

The number of scientific publications addressing wetlands has increased considerably since the early 1970's. "To what end are these wetlands dedicated", one might ask. Or, "Why has there been a sudden interest in the wetlands study?" The purpose of this paper is to answer these questions.

Wetlands and ox-bow lakes are shallow natural depressions with stagnant water rich in organic nutrients and abounding in biodiversity. Many such wetlands have usually developed in chronically floodprone plains of Uttar Pradesh, North Bihar, West Bengal and Assam. It cannot be doubted that wetlands play a vital role in maintaining the ecological balance of the region concerned. In fact, the history and geography of the region would have been very different and certainly very sad but for these wetlands. Although some ambitious projects had been undertaken for the reclamation of some of the wetlands in India, but dominance of economic considerations in these efforts could hardly result in any satisfactory progress.

"Areas of marsh, fen, peatland or water, whether natural or artificial, permanent or temporary, with water that is static or flowing, fresh, brackish or salt, including areas of marine waters, the depth of which at low tide does not exceed six meters" are defined as wetlands according to the Ramsar convention of the IUCN (in Ramsar, Iran, 1971).

After the Ramsar (Iran) convention on Wetlands in 1971 the international community became conscious of the ecodevelopment of these unique ecological systems and today a of number of Indian and international organisations such as the Bombay Natural History Society (BNHS), the International Water Fowl Research Bureau (IWRB), the International Union for Conservation of Nature and Natural Resources (IUCN), the World Wildlife Fund (WWF), etc, have been identifying the wetlands, suggesting conservation technologies and educating people regarding the needs, uses and importance of conservation of wetlands and their fauna and flora. (Asthana, 1979; Cowardin *et al.*, 1979; Yadav and Varshney, 1982; Anonymous, 1990, 1992; Chatrath, 1992).

Unfortunately, the exploration and conservation of biodiversities of wetlands have remained largely neglected in Bihar. Excepting some perfunctory jobs on the eco-development of *Kawar Lake* no significant work has been done on the wetlands of this state. The present study deals with the biodiversitites and reclamation of Baraila lake (a wetland) of Bihar state in India.

The standard methods were used to study various parameters of Baraila lake (A.P.H.A., 1980; Welch, 1952; Cowardin *et al.*, 1979).

Baraila Chaur (Lake/Wetland) is a vast open eutrophic wetland located in Vaishali district of Bihar. It is about 50 km from Patna and 30 km from Hajipur and is surrounded by the rivers Gandak, Noon, Burhi Gandak and Ganga.

Originally Baraila Lake was a vast piece of wetland covered with water. Its status changed in the early 20th century due to construction of railway tracks between Hajipur and Bachhwara and the formation of roadways and Nilhi Kothi (the house where English indigo planters lived). Then the vast area of water was divided into two main parts: one with the present

Baraila Lake and many other chaurs having water bodies of different sizes along with the plant and animal life. During this process a number of uplands were formed which are now inhabited by many villages.

Baraila Lake provides a dazzling nest for a plethora of floral and faunal species which form important trophic components of ecological pyramids. Reclamation and eco-development of Baraila lake would involve preservation of biodiversities, restoration of genetic pool and maintenance of dynamic bio-equilibria that go to make the whole ecosphere stable and sustainable. Not only this; the development of this lake will help in the preservation of some rare species which happen to be important ecological links in the manifold facets of life. Their loss means loss of species and this amounts to erosion of biological wealth of our planet.

Baraila Lake is a vast reservoir of large number of fishes. Ichthyofauna includes *Catla catla, Labeo bata* (Bata fish), *Labeo calbasu* (Calbasu), *Labeo rohita* (Rohu), *Notopterus notopterus* (Moi) *Puntius ticto, Puntius sophore, Puntius chrysopterus* (all three are commonly called Pothia), *Oxygaster bacaila* (Chelwa), *Wallago attu* (Boari), *Mystus tengara* (Tengra), etc.

A number of live fishes are also present in Baraila Lake. These fishes remain alive for sometime outside water due to the presence of air breathing organs in addition to gills. They include *Channa gachua, Channa punctatus, Channa striatus* (all are called Garai), *Clarias bactrachus* (Singhi), *Chanda nama, Chanda ranga* (called Chanda), *Anabas testudineus* (Koi) and *Clarias batrachus* (Mangur).

The fishermen of the adjoining villages carry out fishing practices in this lake and are solely dependent on it for their invelihood.

Baraila lake is also a good receptacle of reeds, grasses and other produce which go in for domestic consumption of the villagers of this locality. The lake produces valuable crops in the form of frogs, lizards, snakes, snails and other forms of life. The lake is rich in invertebrate fauna like fresh water crabs, prawns, water insects and molluscs (including different types of snails and bivalvia).

The lake is a favourite growing ground for plants like *Azolla* and water hyacinth which are natural biological filters and toxin arresters. *Azolla* can treat waste water discharged from sugar and rubber processing industries. Water hyacinth can be used as tertiary pollution filter. Both of these palnts yield good class of biofertilizer. Water hyacinth can also be used for producing biogas.

Baraila wetland stores and absorbs the flood water and thus largely reduces the scale of devastation caused by floods. It plays a vital role in arresting soil erosion controlling flood.

The sub-emergent, emergent and floating vegetation of Baraila is attracting thousands of resident water birds to nest every year, and over millions of migratory birds during winter from far distant temperate regions of *Urasia*. It is one of the largest wintering grounds of the migratory birds in India. Some of the major resident and migratory birds are enlisted below:

LIST A : MIGRATORY BIRDS

S. No.	Scientific Name	Common Name
1.	*Anser anser*	Grey lag goose
2.	*Anser indicus*	Bar-headed goose
3.	*Anas acuta*	Pintail
4.	*Anas crecca*	Common teal
5.	*Anas poecilorhyacha*	Spot bill duck
6.	*Anas platyrhyachos*	Mallard
7.	*Netta rurina*	Red crested poachard
8.	*Aythya ferina*	Common poachard
9.	*Aythya nyroca*	White-eyed poachard
10.	*Aythya fuligula*	Tufted duck
11.	*Aythya marila*	Scaup duck
12.	*Circus aeruginosus*	Marsh Harrier
13.	*Falco peregrinus*	Falcon
14.	*Fulica atra*	Coot
15.	*Larus brunnicephalus*	Brown-headed gull
16.	*Asio flammeus*	Short-cared owl

LIST B : MIGRATORY BIRDS

S. No.	Scientific Name	Common Name
1.	*Phalcrocorax fuscicollis*	Indian shag
2.	*Ardea cinerea*	Grey heron
3.	*Ardea purpurea*	Purple heron
4.	*Ardeola grayii*	Indian pond heron
5.	*Bubulcus ibis*	Cattle erget
6.	*Ardea alba*	Large erget

contd...

S. No.	Scientific Name	Common Name
7.	*Egretta qularis*	Indian reef heron
8.	*Ciconia episcopus*	White-necked Stock
9.	*Dendrocygna bicolor*	Large whistling teal
10.	*Dendrocygna javanica*	Lessor whistling teal
11.	*Sarkidiornis melanotos*	Comb duck
12.	*Elanus caeruleus*	Black-winged kite
13.	*Milvus migrans*	Pariah kite

Different types of angles, vultures, cocks, lapwing, pigeons, doves, keets, cuckoos and swifts have also been seen in and around Baraila Laka. Mammals like nilgai, jackal, fox, mongooses, porcupines, hares, etc., are also present.

CONCLUSION

Based on my personal observations of this wetland for the last three decades as a resident of a nearby village, I am confident that this lake will prove to be a "natural laboratory" for the students of biological and environmental sciences just as that of the Caddo Lake of U.S.A. Caddo Lake is situated at Texas-Louisiana border where students from the local colleges work with faculty members and government agencies to complete research papers of creative projects related to the ecosystem of Caddo lake, one of the largest natural lakes in the South U.S.A.

The Dal Lake of Srinagar is shrinking and turning red owing to an "euglenoid infection" caused by red algae and encroachment by terrorists. A time will come when the Baraila Lake will also die by the encroachment of local people in the catchment areas. It is disappointing to note that neither the State Government nor the Central Government nor any of the state universities has cared to know about this lake which may prove a great asset for the academicians, scientists, naturalists and tourists not only of India but of the whole world.

ACKNOWLEDGEMENTS

The author is thankful to U.S. Department of Interior, Fish and WL Service for providing valuable information. Thanks are also due to Prof. Dr. S. N. Ashan, Pro-Vice Chancellor and Ex-Head of the Department of Zoology, Patna University, Patna- 5 for encouragement.

REFERENCES

APHA (1980). Standard methods for examination of water and waste water. 15th ed. American Pub. Health Association. Washington D.C.

Anonymous (1990). Wetlands of India – A Directory. Govt. of India, Ministry of Environment & Forests, New Delhi.

Anonymous (1992). Conservation of Wetlands in India. Govt. of India, Ministry of Environment & Forests, New Delhi.

Asthana, V. (1979). Limnological studies of lake Chilka, Orissa. Technical Report of MAB project. Dept. of Science & Technology, New Delhi.

Chatrath, K.J.S. (1992). Wetland of India. Ashish Publishing House, New Delhi.

Cowardin, L.M., Carter, V., Golet, F.C. & LaRoe, E. (1979). Classification of wetlands and deepwater habitats of the United States. Biological Services programme. FWS/OBS-97/31.

Welch, P.S. (1952). Limnology. Mc Graw Hill Book Co. 2nd Ed. New York.

Yadav, P. and Varshney, C.K. (1982). Notes on the ecological and socio-economic importance of wetlands of north-eastern India, In : Wetlands-ecology and management. (Gopal, B., Turner, R.E., Wetzel, R.G. and Whigham, D.F. eds.) pp. 149-151. National Institute of Ecology and International Scientific Publication, New Delhi.

8

Seasonality on the Egg Production and the Energetics of Egg Development in Macrobrachium lamarrei

P. Maria Charles

INTRODUCTION

Though, there is considerable information on the occurrence of *Macrobrachium lamarrei* in South India (Raman *et al.*, 1978; Holthuis, 1980), work on its breeding biology and egg development is meagre. It is known that the optimum egg size and number of the brood maximize the survival and growth of the larvae, particularly in aquatic organisms whose parents do not nourish the young ones (Blaxter and Hempel, 1963; Begenal, 1969; Gall, 1974; Svardson, 1948; Lack, 1954). In the present study, we report on the seasonality in the egg production and energetics of egg development. Since, environmental temperature plays a major role on the crustacean egg development (Berry, 1971; Herring, 1974; Hart, 1980) the influence of water temperature on the rate of egg development has also been investigated.

MATERIALS AND METHODS

Berried specimens of *Macrobrachium lamarrei* collected fortnightly between June 1992 and May 1993 from Chembarambakkam pond, Madras were used. For egg count, in each collection, eggs were stripped off from the pleopod onto a glass slide and enumerated directly about 50 ovigerous females, after measuring the length of the prawn. As egg loss has been observed in older eggs, only females carrying dark green eggs (freshly laid) were used for fecundity study. The size of the egg was determined using an ocular micrometer.

For studying the fecundity of the laboratory-reared prawns, vitellogenic females procured from the pond were reared in the density of 2♂ and 5 to 8♀ per aquaria of 70 cm × 40 cm × 40 cm size in the laboratory during the period from June 1992 to May 1993. The prawns were held under normal photoperiod (12 D; 12 L) and water temperature (29 ± 2 °C). Prawns were fed with live tubifex worm *ad libitum*; aquaria were cleaned and water was changed every alternate day. The number of eggs in each clutch was enumerated within a day of spawning and length of the mother was measured.

Bimonthly estimations of biochemical components of the egg were made. The water content was calculated as the difference between the wet weight and dry weight of the eggs. Protein content of the eggs was estimated by Folinciocalteu method (Lowry *et al.*, 1951). Carbohydrate was estimated spectrophotometrically using Anthrone reagent (Roe, 1955). For the estimation of lipid, the method of Folch *et al.* (1957) was employed and energy value was determined using a Parr 1412 semi-micro bomb calorimeter. The body length and clutch size of the mother prawns were tested using regression method.

RESULTS

Seasonality in egg production

Wild population

Monthly egg production is estimated between June 1992 and May 1993 in the broods of the females collected from the habitat. In all the months there is an increase in the number of eggs laid with the increase in the size of the mother (Fig. 1). The mean clutch size computed for all the berried prawns from the smallest of 40 mm to the biggest size of 56 mm ranges between 55 and 111 eggs (Table 1). The same analysed on monthly basis varied significantly in different months – increases from 71 eggs in June to 112 eggs in November and then decreases to 62 eggs in May. However, fecundity was inverse during this period when analysed on unit weight basis (Table 2). Hence, the maximum number of eggs per clutch has been recorded in the month of November when peak reproductive activity was observed in the population.

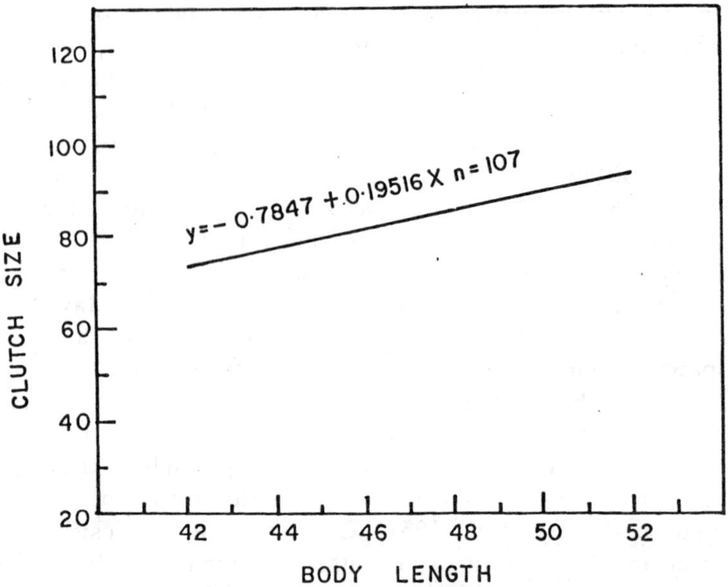

FIG. 1: REGRESSION LINE SHOWING THE RELATIONSHIP BETWEEN THE BODY LENGTH AND CLUTCH SIZE OF *M. LAMARREI* COLLECTED FROM THE HABITAT

LABORATORY REARED PRAWNS

Fecundity estimation made on laboratory reared prawns during the same period also showed a positive correlation between the clutch size and the size of the mother (Fig. 2) and that the berried prawns ranging in length between 40 to 56 mm laid a mean minimum and maximum of 55 and 109 eggs respectively with an average of 80 eggs (Table 3).

Although there is difference in the mean monthly egg production of the wild and laboratory reared prawns, maximum and minimum clutch size occur in both the groups in the month of November and May respectively (Fig. 3). The monthly variations in the clutch size tested with analysis of variance are found to be statistically significant for all the observed size classes of wild and laboratory reared prawns (Table 4).

EGG SIZE AND WEIGHT

The mean size of the egg determined during this study period which varied significantly between months – increasing

Table 1: MONTHLY VARIATION IN THE FECUNDITY OF *M. LAMARREI* COLLECTED FROM THE HABITAT

Size mm	1992							1993					Mean
	June	July	Aug.	Sept.	Oct.	Nov.	Dec.	Jan.	Feb.	March	Apr.	May	
40	–	31 ± 5	58 ± 1	–	–	–	–	–	–	–	64 ± 5	65 ± 2	55 ± 16
42	–	34 ± 5	70 ± 2	65 ± 3	68 ± 2	–	99 ± 2	92 ± 5	–	–	54 ± 2	66 ± 3	68 ± 20
44	71 ± 2	95 ± 5	71 ± 3	74 ± 5	97 ± 2	76 ± 2	74 ± 3	102 ± 5	68 ± 1	60 ± 6	50 ± 6	–	76 ± 16
46	70 ± 4	82 ± 2	76 ± 1	80 ± 6	104 ± 2	107 ± 1	108 ± 8	74 ± 5	75 ± 5	70 ± 7	–	62 ± 3	82 ± 16
48	72 ± 3	90 ± 3	114 ± 5	112 ± 2	119 ± 5	115 ± 1	107 ± 9	94 ± 2	76 ± 1	78 ± 8	66 ± 5	–	94 ± 20
50	–	–	100 ± 1	105 ± 5	105 ± 5	107 ± 5	103 ± 2	82 ± 2	44 ± 2	60 ± 4	–	54 ± 4	84 ± 25
52	–	–	110 ± 4	104 ± 1	83 ± 6	126 ± 3	129 ± 3	102 ± 1	76 ± 3	102 ± 6	63 ± 3	± 2	99 ± 22
54	–	–	–	–	–	129 ± 7	–	–	–	90 ± 5	73 ± 5	–	97 ± 29
56	–	110 ± 1	–	120 ± 4	107 ± 2	124 ± 5	115 ± 4	110 ± 3	117 ± 6	100 ± 3	90 ± 4	–	111 ± 10
Mean	71 ± 1	74 ± 30	86 ± 20	94 ± 21	98 ± 17	112 ± 18	105 ± 17	94 ± 13	76 ± 24	80 ± 18	66 ± 13	62 ± 5	85

TABLE 2 : MONTHLY VARIATION IN THE CLUTCH SIZE OF *M. LAMARREI* (EGG NUMBER PER UNIT WEIGHT; 100 MG)

Month	Number of observations	Mean egg number
1992		
June	11	160 ± 2.1825
July	7	160 ± 1.8257
August	7	153 ± 2.8702
September	9	146 ± 2.0616
October	6	140 ± 1.7512
November	7	133 ± 2.9277
December	8	143 ± 1.8516
1993		
January	5	143 ± 1.8708
February	5	143 ± 2.2361
March	4	146 ± 1.7078
April	4	150 ± 1.2583
May	5	153 ± 2.2361

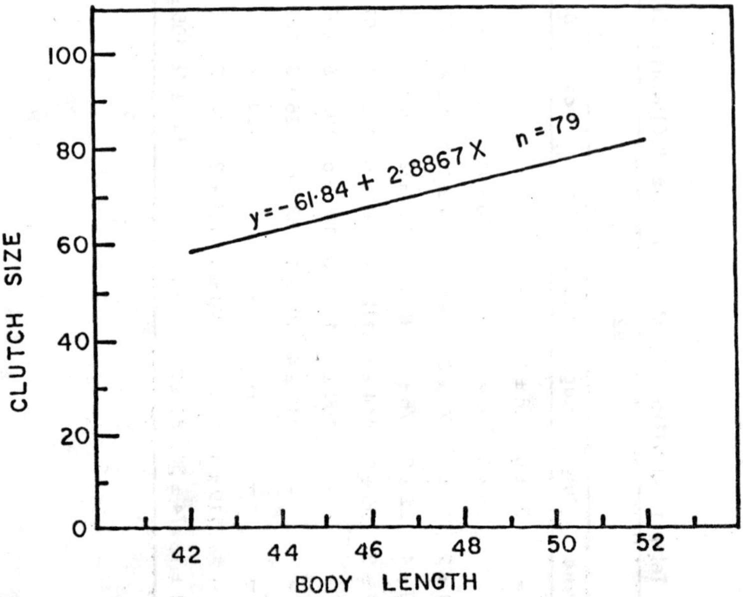

FIG. 2 : REGRESSION LINE SHOWING THE RELATIONSHIP BETWEEN THE BODY LENGTH AND CLUTCH SIZE OF *M. LAMARREI* REARED IN THE LABORATORY

Table 3 : MONTHLY VARIATION IN THE FECUNDITY OF *M. LAMARREI* REARED IN THE LABORATORY

Total length	1992							1993					Mean
mm	June	July	Aug.	Sept.	Oct.	Nov.	Dec.	Jan.	Feb.	March	Apr.	May	
40	46 ± 8	24 ± 14	74 ± 19	25 ± 18	84 ± 10	48 ± 15	50 ± 5	85 ± 13	67 ± 11	45 ± 13	55 ± 11	53 ± 5	55±19.92
42	43 ± 10	27 ± 22	60 ± 11	25 ± 17	63 ± 20	70 ± 9	57 ± 9	87 ± 15	62 ± 8	49 ± 12	54 ± 15	58 ± 12	55±17.22
44	85 ± 11	85 ± 20	76 ± 9	90 ± 20	81 ± 11	71 ± 9	64 ± 12	73 ± 19	66 ± 9	41 ± 12	40 ± 21	46 ± 8	68±17.44
46	76 ± 10	72 ± 9	71 ± 16	81 ± 13	63 ± 19	76 ± 9	74 ± 6	72 ± 20	54 ± 5	75 ± 11	74 ± 5	57 ± 6	70±8.15
48	81 ± 8	80 ± 8	65 ± 19	91 ± 10	70 ± 14	115 ± 21	115 ± 21	86 ± 18	68 ± 8	79 ± 15	56 ± 9	63 ± 6	81±18.92
50	72 ± 5	93 ± 14	72 ± 20	91 ± 8	79 ± 12	105 ± 12	97 ± 15	51 ± 12	86 ± 8	79 ± 19	82 ± 12	66 ± 9	81±14.76
52	68 ± 6	97 ± 13	67 ± 5	92 ± 9	81 ± 8	124 ± 21	122 ± 18	124 ± 14	107 ± 12	107 ± 4	74 ± 11	68 ± 9	94±22.53
54	96 ± 9	96 ± 19	95 ± 9	94 ± 5	116 ± 14	150 ± 18	157 ± 18	127 ± 15	125 ± 14	95 ± 5	75 ± 9	66 ± 10	108±27.88
56	95 ± 10	101 ± 7	102 ± 13	94 ± 20	116 ± 16	151 ± 17	160 ± 13	124 ± 18	113 ± 11	99 ± 5	81 ± 5	74 ± 12	109±25.78
Mean	74 ± 18.98	75 ± 29.5	76 ± 13.87	76 ± 29.11	84 ± 19.89	101 ± 37.05	100 ± 41.67	92 ± 26.98	83 ± 25.74	74 ± 24.39	66 ± 14.71	61 ± 8.56	80

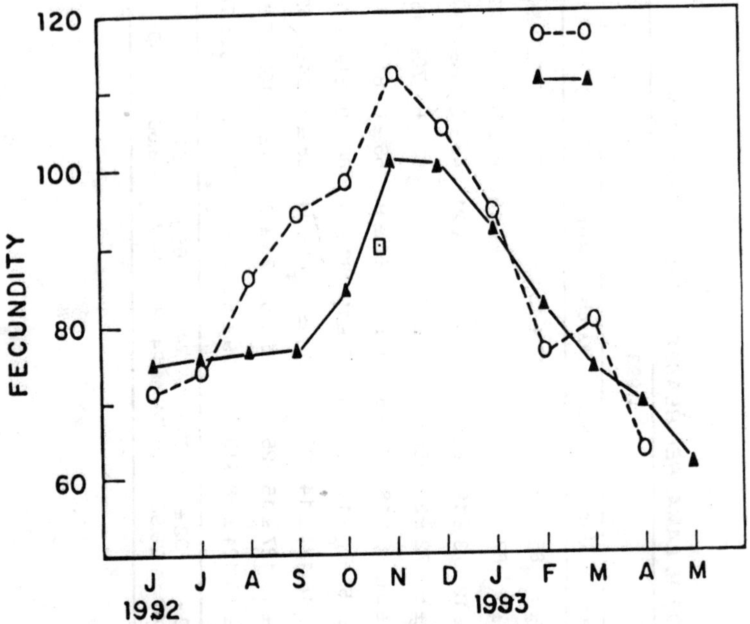

FIG. 3 : MONTHLY VARIATIONS IN THE FECUNDITY OF FIELD COLLECTED AND LABORATORY REARED *M. LAMARREI* OF THE YEAR 1992-93 (O...O COLLECTED FROM THE HABITAT; Δ—Δ LABORATORY REARED).

TABLE 4 : ANALYSIS OF VARIANCE TO TEST THE SIGNIFICANCE OF MONTHLY VARIATIONS IN CLUTCH SIZE VS BODY LENGTH

Source of Variance	SS	DF (n–1)	Mean squares SS/DF	F = P/Q	Fe
Between months	15203.139	11	1382.1035	5.01158	2.456
Between size	40086.333	8	5010.7916	18.169	2.718
Error	24268.778	88	275.78157		
Total	**79558.25**	**107**			

RESULT:

(i) Since the calculated F value 5.01158 is greater than the expected F value 2.456 there is significant difference in the clutches between months.

(ii) Since the calculated F value 18.169 is greater than the expected F value 2.718 there is significant different in the clutches between sizes.

from June to November and decreasing in May – suggests that individual egg size is also maximised during peak reproductive months (Table 5). Correspondingly, the monthly mean weight of wet and dry egg given in Table 6 indicates that the highest value of wet and dry weight of a single egg is found in November; while the lowest value is noticed in the month of June.

TABLE 5 : MONTHLY VARIATION IN THE EGG SIZE OF *M. LAMARREI*

Month	Number of observations	Mean length of egg (μm)	Mean width (μm)	Average
1992				
June	5	1219 ± 1.5811	1058 ± 3.8079	1038.5
July	5	1219 ± 2.8284	1081 ± 2.6458	1150.0
August	5	1311 ± 2.0000	1035 ± 1.2247	1173.0
September	5	1334 ± 1.5811	1035 ± 1.5811	1185.0
October	5	1426 ± 3.6742	1104 ± 2.6456	1265.0
November	5	1449 ± 2.0000	1104 ± 2.9154	1276.5
December	5	1403 ± 2.9154	1035 ± 3.3166	1219.0
1993				
January	5	1380 ± 1.5811	1127 ± 5.1478	1253.5
February	5	1403 ± 2.0000	1030 ± 1.5811	1219.0
March	5	1334 ± 1.5811	1104 ± 3.5355	1219.0
April	5	1311 ± 2.3452	1035 ± 1.5811	1173.0
May	5	1242 ± 2.5495	1058 ± 2.7386	1150.5

BIOCHEMICAL COMPONENTS OF THE EGG

The total organic components namely, protein, carbohydrate and lipid of a single egg varied in different months, with an increase from June to November and a decrease in May (Table 6). Protein increases from 0.095 in June to 0.123 mg in November, thereafter decreases to 0.1073 mg/egg dry wt in May. The lipid also increases from 0.1958 in June to 0.2186 in November, declining afterward to 0.1997 mg in May. The amount of carbohydrate ranges between 0.0052 mg in September and 0.0144 mg in August. The ash content also varies from 0.0102 in June to 0.0122 mg in September but decreases to 0.0082 in May. However, the biochemical components expressed in percentage, do not vary much between months excepting a slight fluctuation in lipid content (Table 7).

Table 6 : MONTHLY FLUCTUATION IN THE BIOCHEMICAL COMPOSITION OF A SINGLE EGG DURING THE YEAR 1992-93 (ALL VALUES IN MG DRY WT; SINGLE EGG AND WATER IN MG LIVE WEIGHT)

Month	Single egg	Water	Dry single egg	Protein	Carbohydrate	Lipid	Ash
1992							
June	0.625 ± 0.0004	0.3140 ± 0.0005	0.3110 ± 0.0009	0.0950 ± 0.0005	0.0100 ± 0.0005	0.1958 ± 0.0008	0.0102 ± 0.0003
July	0.625 ± 0.0005	0.3033 ± 0.0004	0.3217 ± 0.0009	0.0975 ± 0.0007	0.0075 ± 0.0004	0.2076 ± 0.0005	0.0091 ± 0.0009
Aug.	0.653 ± 0.0005	0.3287 ± 0.0005	0.3243 ± 0.0009	0.1058 ± 0.0007	0.0144 ± 0.0006	0.1971 ± 0.0005	0.0070 ± 0.0008
Sept.	0.685±0.0006	0.3461 ± 0.0006	0.3389 ± 0.0004	0.0961 ± 0.0005	0.0052 ± 0.0007	0.2255 ± 0.0004	0.0121 ± 0.0008
Oct.	0.714 ± 0.0004	0.3579 ± 0.0007	0.3561 ± 0.005	0.1221 ± 0.0005	0.0059 ± 0.009	0.2172 ± 0.0003	0.0109 ± 0.0009
Nov.	0.752± 0.0004	0.3913 ± 0.0006	0.3607 ± 0.0006	0.1230 ± 0.0005	0.0069 ± 0.0008	0.2186 ± 0.0005	0.0122 ± 0.0010
Dec.	0.699 ± 0.0005	0.3635 ± 0.0003	0.3355 ± 0.0009	0.1187 ± 0.0010	0.0106 ± 0.0009	0.1992 ± 0.0007	0.0070 ± 0.0030
1993							
Jan.	0.699 ± 0.0009	0.3511 ± 0.0004	0.3479 ± 0.0006	0.1133 ± 0.0010	0.0081 ± 0.0005	0.2184 ± 0.0008	0.0081 ± 0.0005
Feb.	0.699 ± 0.0004	0.3539 ± 0.0005	0.3451 ± 0.0006	0.1095 ± 0.0009	0.0084 ± 0.0005	0.2207 ± 0.0007	0.0064 ± 0.0005
March	0.685 ± 0.0005	0.3477 ± 0.0005	0.3373 ± 0.0007	0.1186 ± 0.0005	0.0133 ± 0.0005	0.1976 ± 0.0004	0.0076 ± 0.0009
April	0.667 ± 0.0003	0.3341 ± 0.0004	0.3329 ± 0.0008	0.1171 ± 0.0006	0.0069 ± 0.0003	0.2027 ± 0.0005	0.0072 ± 0.0009
May	0.654 ± 0.0006	0.3270 ± 0.0005	0.3270 ± 0.0003	0.1073 ± 0.0007	0.0118 ± 0.0007	0.1997 ± 0.0006	0.0082 ± 0.0009

TABLE 7 : MONTHLY CHANGES IN THE BIOCHEMICAL COMPOSITION OF EGG LAID DURING THE YEAR 1992-93

Month	Water (%)	Biochemical components (%)			Ash (%)
		Protein	Carbohydrate	Lipid	
1992					
June	50.24	15.20	1.60	31.33	1.63
July	48.53	15.60	1.20	33.21	1.46
August	50.34	16.20	2.20	30.18	1.08
September	50.53	14.04	0.74	32.93	1.76
October	50.12	17.10	0.82	30.44	1.52
November	52.03	16.36	0.92	29.07	1.62
December	52.00	16.98	1.52	28.50	1.00
1993					
January	49.77	16.36	1.17	31.52	1.18
February	50.63	15.66	1.30	31.57	0.93
March	50.76	17.32	1.94	28.84	1.14
April	50.09	17.55	0.89	30.41	1.09
May	50.00	16.40	1.80	30.54	1.26

ENERGY VALUE OF THE EGG

Energy value of the egg expressed in calories per gram dry weight rises from 6056 in June to 6851 in November and drops to 5643 in May. Similarly, when expressed on per egg basis, this value increases from 1.8740 in June to 2.4715 in November and declines to 1.8453 cal/egg wt in May (Table 8).

EFFECTS OF TEMPERATURE ON EGG DEVELOPMENT

To study the effect of temperature on egg development, 5 sets of 7 to 10 ovigerous females bearing 0 day old eggs collected from the pond were brought to the laboratory and reared in aquaria of 70 cm × 40 cm × 40 cm size in four selected months with different temperature regimes; December 1992 (27.8°C), March 1993 (28.9°C), September 1992 (30.6°C) and May 1993 (31.7°C).

The results of the experiment indicate that at different temperatures the duration of egg incubation varied from 16 to 23 days (Table 9). The mean number of days taken for the complete egg development in all the different temperatures is 19.9. Hence for a temperature difference of 3.9°C a difference of

7 days is evident. In other words, for every 1°C increase of temperature the egg developmental time declines by 1.8 days.

TABLE 8 : MONTHLY VARIATION IN THE ENERGY VALUE OF EGGS LAID DURING THE YEAR 1992-93

Month	Energy value (cal / g dry wt)	Single egg (cal / egg dry wt)
1992		
June	6056 ± 12	1.8740 ± 0.0007
July	6318 ± 25	2.0328 ± 0.0008
August	6352 ± 16	2.0597 ± 0.0005
September	6382 ± 08	2.0698 ± 0.0011
October	6506 ± 10	2.3170 ± 0.0003
November	6851 ± 13	2.4715 ± 0.0009
December	6512 ± 28	2.3845 ± 0.0008
1993		
January	6382 ± 24	2.2205 ± 0.0009
February	6055 ± 15	2.1894 ± 0.0006
March	5951 ± 19	2.1000 ± 0.0012
April	5865 ± 02	1.9523 ± 0.0005
May	5643 ± 08	1.8453 ± 0.0006

ENERGY COST OF EGG INCUBATION

The mean energy contents of a fresh egg and egg before hatching at different seasons are 2.0999 and 1.2916 calories respectively. The mean value of the energy utilized for embryonic development is 0.8084 cal/egg. It is obvious that for a mean development period of 19.9 days an egg needs 0.8084 calories of energy. For one day, an egg of *M. lamarrei* utilizes 0.0404 calories of energy. The initial energy value of the freshly spawned eggs, energy utilized during egg incubation and energy left over with the larvae at hatching in selected months at various temperature regimes are given in Table 10.

The results indicate that the egg laid in December (27.8°C) possessing 2.3845 calories of initial energy hatches out the larva with the energy value of 1.4093 calories. At the same time the egg laid in May (31.7°C) possessing 1.8453 calories of initial energy hatches out into larva holding 1.1669 calories. In other words, the larva hatched in December at 27.8°C temperature holds 0.2424 calories over the larva released in May at 31.7°C.

Table 9 : SEASONALITY ON THE TEMPERATURE EFFECT OF EGG DEVELOPMENT IN *M. LAMARREI*

Month	Temperature °C	Development duration between stages (in days)					Total Duration (in days)
		I to II	II to III	III to IV	IV to V	V to VI	
December 1992	27.8	6.60 ± 0.5	5.60 ± 0.0	4.80 ± 0.0	3.60 ± 0.0	2.40 ± 0.0	23
March 1993	28.7	6.00 ± 0.5	6.00 ± 0.5	4.40 ± 0.0	3.00 ± 0.0	2.00 ± 0.0	21
September 1982	30.6	5.75 ± 0.5	4.75 ± 0.0	3.50 ± 0.0	2.75 ± 0.0	1.25 ± 0.0	18
May 1993	31.7	5.6 ± 1.0	4.00 ± 0.0	3.00 ± 0.5	2.33 ± 0.0	1.33 ± 0.0	16
Mean	–	**5.90 ± 0.44**	**5.30 ± 0.53**	**3.93 ± 0.82**	**2.92 ± 0.53**	**1.75 ± 0.55**	**19.9**

Table 10 : SEASONALITY IN THE ENERGY COST OF EGG INCUBATION IN *M. LAMARREI*

Month	Temperature °C	Energy value (cql / egg or larva dry wt)		
		Newly spawned egg	Freshly hatched larva	During incubation
December 1992	27.8	2.3845 ± 0.0008	1.4093 ± 0.0007	0.9752 ± 0.0002
March 1993	28.9	2.1000 ± 0.0005	1.3000 ± 0.0006	0.8000 ± 0.0003
September 1992	30.6	2.0698 ± 0.0006	1.2900 ± 0.0005	0.7656 ± 0.0004
May 1993	31.7	1.8453 ± 0.0009	1.1669 ± 0.0003	0.6784 ± 0.0005
Mean	–	**2.0999 ± 0.2211**	**1.2916 ± 0.0991**	**0.8084 ± 0.1245**

Clearly, for a difference of 3.9°C the egg spends 0.2424 calories of energy more. Therefore, for 1°C increase of temperature during the development the egg takes 0.0622 calories of energy more.

DISCUSSION

The size as well as the number of eggs spawned vary widely among different natantian decapods (Bliss, 1968; Shokita, 1973; Nishino, 1980; Mashiko, 1982a, 1982b, 1983). The number of eggs in a brood depends on the size of the mother prawn and a linear relationship has been obtained between the body length and clutch size in crustaceans (Farmer, 1974; Fish, 1975; Balasundaram and Pandian, 1982). However, in species where such relationship is not apparent, a correlation is established between the length of the ripe ovary and the size of the mother (Balasundaram and Pandian, 1982; Charles, 1984). Fecundity data obtained in the present study, though in agreement with previous works, show a marked variation between the spawn of field collected and laboratory reared groups.

Marine crustacean as well as the forms that return to sea for breeding bear very large number of small eggs (Bliss, 1968; Mashiko, 1983). On the contrary, species that return to or remain in fresh water for spawning lay only a small number of large yolky eggs (Shokita, 1973; Gangotri *et al.*, 1977; Pillai, 1981). This is because the eggs of marine crustaceans develop within a relatively shorter period, thereby releasing the young ones at very early larval stages such as zoea or protozoea; whereas the freshwater forms as well as terrestrial species incubate the eggs for longer periods, thereby releasing the young ones at an advanced stage (Gangotri *et al.*, 1977; Pillai, 1981). In general, the marine crustaceans tend to breed repetitively holding short period of gametogenesis whereby producing quick broods within its breeding season. Naturally, forms that are producing heavily yolk-laiden eggs take longer period for vitellogenesis. This is supported by the present study on *M. lamarrei* where the number of eggs are less compared to marine penaeid prawns or the freshwater prawns which migrates to the estuary for spawning. This is also suggested in some freshwater fishes (Goto, 1975; Mizuno and Niwa, 1961; Mizuno, 1960).

Although the adaptive importance of this has been very well correlated with trophic conditions of the medium for the better survival of the young ones by several works (Hutchinson, 1951; Marshall, 1953; Lack, 1954), Thorson (1950) and Mashiko (1983) substantiated this with salinity condition of the medium and claimed that the large egg and larva of freshwater prawn is an adaptiveness to increased energy demand for osmoregulation in freshwaters. Contradicting this, spawning of many small eggs in the upper reaches and laying of fewer but larger eggs in the lower courses of the river has also been reported in *M. nipponense* (Mashiko, 1983) and this is correlated with the velocity of water current which is high at the upper reaches and almost absent in the lower courses.

The size and number of eggs determined in different seasons of the present study indicate that there occurs seasonality in egg production. The biggest and maximum number of eggs in the brood are observed to spawn in November while the smallest and minimum are in May. The wet and dry weight of a single egg laid in different months analysed indicate that bigger eggs with maximum organic matter (like protein, carbohydrate and lipid) and ash are obtained in the month of November when the maximum rainfall and best trophic conditions of the pond are recorded (Charles, 1984). Incidentally, the optimum environmental conditions (such as dissolved oxygen, pH, temperature) nutrient content and primary productivity of the rainy season that facilitated the abundance of food material in the medium correlated favourably with the increased organic content and energy level of the eggs laid in November.

The temperature effect of egg incubation studied in decapods suggests that there is an inverse relationship between these two (Berry, 1971; Perkins, 1972; Herring, 1974; Hart, 1980). In the present study, the temperature recorded in the medium fluctuated between 27.8°C and 31.7°C with a difference of 3.9°C. The observed incubation duration of 23 days in December at 27.8°C decreasing to 16 days in May at 31.7°C temperature suggests that the development rate of egg spawned in summer is significantly faster than that of the eggs in winter. The decline of egg developmental time by 1.8 days for every 1°C temperature increase of the medium of this study is well comparable with the egg development rate of other natantians (Table 11).

TABLE 11 : TEMPERATURE AND EGG DEVELOPMENT DURATION OF OTHER NATANTIANS

Species	Temperature (°C)	Developmental duration (in days)	Reference
Macrobrachium rosenbergii	26–28	19–21	Ling, 1969
Macrobrachium acanthurus	23–27	16–18	Choudhury, 1971
Macrobrachium amazonicum	24–30	24–19	Guest, 1979
Macrobrachium nobilii	22–32	15–8	Balasundaram, 1980
Macrobrachium heterochirus	26–28	19–15	Ching and Velez, 1985
Macrobrachium lamarrei	28–32	23–16	Present study, 1993

Interestingly, this temperature dependence of egg development has direct impact on the size and yolk energy utilization pattern of the eggs laid in different seasons. The large eggs laid in winter season (December) holding greater energy content (2.3845 cal/egg) take longer incubation time at low temperature (27.8°C); whereas eggs spawned in summer (May) possessing less energy (1.8453 cal/egg) incubate faster due to high temperature (31.7°C). Similarly, it is estimated that there is significant difference in the amount of yolk energy left over with the larvae released in the months of December and May. At hatching, an estimated amount of 0.2424 calories of yolk energy is seen more with the larvae released in December over May. This may be adaptive significance to the larvae released in December which has to survive longer than the ones released in May. It is also noteworthy to mention from the present study that the fresh-water prawn *M. lamarrei* lays large many eggs repetitively in the rainy season as the medium is enriched with nutrient and conducive for the mother to lay more eggs and the larvae to get dispersed like few other freshwater decapods (Gangotri *et al.*, 1977; Pillai 1981); whereas its marine counterparts, lay large number of eggs during summer months as they do not suffer scarcity of water.

SUMMARY

Macrobrachium lamarrei (H. Milne Edwards) is a non migratory, caridian prawn inhabiting the freshwater bodies in South India. Being a continuous and repetitive breeder, it lays about hundreds of eggs and reproduces once in 25 days to a maximum of three times at favourable season. Eggs laid after the pre-spawning moult is brooded for about 20 days in the abdominal pleopods until hatching.

A positive relationship is established between the number of eggs spawned and the size of the mother. The size and number of eggs spawned per brood by a female vary significantly between months. The weight of a wet and dry egg differs seasonally and is maximum in November. However, biochemical components of the egg expressed in percentage dry weight do not vary much between months, excepting a slight fluctuation in lipid content.

This seasonality in the egg production is correlated with the trophic conditions of the medium. The duration of egg incubation at different temperatures vary between 16 and 23 days and declines by 1.8 days for every 1°C temperature increase. The significance of the results are discussed in relation to larval energetics.

ACKNOWLEDGEMENTS

The authors thank Prof. K. Ramalingam for providing facilities and encouragement. This research was conducted with assistance, which PMC acknowledges, from the Council of Scientific and Industrial Research, New Delhi. He is also indebted to Prof. T. Subramanium for his interest and help.

REFERENCES

Balasundaram, C. (1980), Ecophysiological studies in prawn culture (*Macrobrachium nobilii*). Ph.D. thesis, Madurai Kamaraj University, Madurai, 213 pp.

Balasundaram, C. and T.J. Pandian (1982), Egg loss during incubation in *Macrobrachium nobilii* (Henderson and Mathai). J. Exp. Mar. Biol. Ecol., **59**: 289-299.

Begenal, T.B. (1969), Relation between egg size and fry survival in brown trout *Salmo trutta* L.J. Fish. Biol. **1**: 349-353.

Berry, P.F. (1971), The biology of spiny lobster *Panulirus homarus* (Linnaeus) of the East Coast of Southern Africa. Invest. Rep. Oceanogr. Res. Inst., **28**: 1-75.

Blaxter, J.H.S. and G. Hempel (1963), The influence of egg size on herring larvae (*Clupea harengus L.*) J. Cons., **28**: 211-240.

Bliss, D.E. (1968), Transition from water to land in decapod crustaceans. Amer. Zool., **8**: 355-392.

Charles, P.M. (1984) Reproductive physiology of a freshwater prawn *Macrobrachium lamarrei* (H. Milne Edwards) Ph.D. thesis, University of Madras, Madras, 106 pp.

Ching, C.A. and M.J. Velez (1985), Mating, incubation and embryo number in the freshwater prawn *Macrobrachium heterochirus* (Wiegmann, 1836) (Decapoda, Palaemonidae) under laboratory conditions. Crustaceana **49**: 42-48.

Choudhury, P.C. (1971), Laboratory rearing of larvae of the palaemonid shrimp *Macrobrachium acanthurus* (Wiegmann, 1836) Crustaceana, **21**: 113-125.

Farmer, A.S.D. (1974), Reproduction in *Nephrops norvegicus* (Decapoda: Nephropidae) J. Zool. Lond. **174**: 161-183.

Fish, J.D. (1975), Development, hatching and brood size in *Bathyporeia pilosa* and *B. pelagica* (Crustacea: Amphipoda) J. Mar. Biol. Assoc. U.K., **59**: 355-368.

Folch, J., Lees, M. and G.H. Sloane-Stanley (1957), A simple method for the isolation and purification of total lipids from animal tissues. J. Biol. Chem., **266**: 497-509.

Gall, G.A.E. (1974), Influence of size of egg and age of female on hatchability and growth in rainbow trout. Calif. Fish Game, **60**: 26-35.

Gangotri, M.S., Vasantha, N. and S.A.T. Venkatachari (1977), Sexual maturity and breeding behaviour in the fresh-water crab, *Barytelphusa querini* Milne Edwards. Proc. Indian Acad. Sci., **87**: 195-201.

Goto, A. (1975), Ecological and morphological divergence of freshwater sculpin, *Cottus nozawae* Snyder I. spawning behaviour and process of the development in the post hatching stage. Bull. Fac. Fish. Hokkaido Univ. **26**: 31-37 (In Japanese with English summary).

Guest, W.C. (1979), Laboratory life history of the palaemonid shrimp *Macrobrachium amazonicum* (Heller) (Decapod,

Palaemonidae). Crustaceana, **37**: 141-152.

Hart, R.C. (1980), Embryonic duration and post embryonic growth rates of the tropical freshwater shrimp *Caridina nilotica* (Decapoda: Atyidae) under laboratory and experimental field conditions. Fresh. Biol., **10**: 297-315.

Herring, P.J. (1974), Observations on embryonic development of some deep living decapod crustaceans with particular reference to species of *Acanthephyra*. Mar. Biol., **25**: 25-33.

Holthuis, L.B. (1980), Shrimps and prawns of the world. FAO Fish. Synop., **1**: 261.

Hutchinson, G.E. (1951), Copepodology for the ornithologist. Ecology, **32**: 571-577.

Lack, D. (1954), The natural regulation of animal numbers, Oxford Univ. Press, London, 343 p.

Ling, S.W. (1969), The general biology and development of *Macrobrachium rosenbergii* (de Man). FAO Fish. Rep., **57**: 607-619.

Lowry, D.H. Rosebrough, N.J., Farr. J.L. and R.J. Randall (1951), Protein measurement with the folin phenol reagent. J. Biol. Chem. **193**: 265-275.

Marshall, N.B. (1953), Egg size in arctic, antarctic and deep sea fishes. Evolution, **7**: 328-341.

Mashiko, K. (1982a), Differentiation in the reproductive characteristics of the prawn *Macrobrachium nipponense* (DE HAAN). The heredity (The Iden), **36**(1): 81-87 (In Japanese).

Mashiko, K. (1982b), Differences in both the egg size and the clutch size of the freshwater prawn *Palaemon paucidens* de Haan in the Sagami river, Jap. J. Ecol. **32**: 445-451.

Mashiko, K. (1983), Difference in the egg and clutch size of the prawn *Macrobrachium nipponense* (de Haan) between brackishwater and freshwater of a river. Zool. Mag., **92**: 1-9.

Mizuno, N. (1960), Study on a freshwater goby, *Rhinogobius similis* Gill, with a proposition on the relationship between land locking and speciation of some freshwater gobies in Japan. Mem. Coll. Sci. Univ. Kyoto, Ser. B, **27**: 97-115.

Mizuno, N. and H. Niwa (1961), Two ecological types of the freshwater scorpion *Cottus pollux* Gunter. Zool. Mag. **70**: 267-275 (In Japanese with English summary).

Nishino M. (1980), Geographical variation in body size, brood size and egg size of a freshwater shrimp, *Palaemon paucidens* De Haan with some discussion on brood habit. Jap. J. Limnol., **41**: 185-202.

Perkins, H.C. (1972), Developmentai rates at various temperature of embryos of the northern lobster *Homarus americanus* Milne Edwards. Fish. Bull, U.S. **7**: 59-99.

Pillai, C.K. (1981), Studies on the breeding biology and egg development in a freshwater crab, *Paratelphusa hydrodromous* (Herbst). Ph.D. thesis, University of Madras, 138 p.

Raman, K.V.A., Reddy, S.R., Katre, S. and S. Ayyappan (1978), Occurence and distribution of freshwater prawns in and around Bangalore. Vignana Bharathi, **4**: 78-87.

Roe, J.H. (1955), The determination of sugar in blood and spinal fluid with anthrone reagent. J. Biol. Chem., **20**: 335-343.

Shokita, S. (1973), Abbreviated larval development of the freshwater prawn, *Macrobrachium shokitai* Fujino et Baba (Decapoda, Palaemonidae) from Iriomote Island of the Ryukyus. Annot. Zool. Japon., **46**: 111-126.

Svardson, G. (1948), Natural selection and egg number in fish. Rep. Inst. Freshwat. Res. Drottningholm no. **29**: 115-122.

Thorson, G. (1950), Reproductive and larval ecology of marine bottom invertebrates. Biol. Rev. **25**: 1-45.

9

Oxygen Consumption in Relation to the Treatment of Certain Biogenic Amines in Crab *Barytelphusa Cunicularis*

K.C. Vipradas

INTRODUCTION

The role that endocrine plays in the control of metabolic rate of crustaceans amongst other higher invertebrates has been investigated by several workers. The existence of cortacholaminergic and seratonergic neurons have been reported from eyestalk and ganglia of crustaceans. (Aramant & Elofsson, 1976; Fingerman & Fingerman, 1977). The biogenic amines are known to participate in the metabolic processes of the animals. (Schneider & Macann, 1969). It plays an important role in the metabolism of crustaceans and molluscs. It releases neurohormones in crustaceans. (Fingerman *et. al.* 1974; Fingerman & Fingerman 1977).

Acetylcholine has been found in high concentration in gastropod ganglia (Tank, 1966) and in the ganglion of bivalves (Cottrell, 1966 and Laverack, 1968). Neurons responsive to acetylcholine, 5-hydroxytryptamine, dopamine, noradrenalin, glucamet and histamine have been detected in Molluscan ganglia. Thus taking into consideration the role played by biogenic amines in the metabolic processes of crustaceans (typical example of freshwater crab) and molluscs, the present study has been undertaken to study the effect of two biogenic amines namely acetylcholine and adrenalin on the rate of oxygen consumption of the fresh water crab *Barytelphusa cunicularis.*

MATERIALS AND METHODS

Freshwater crab *B. cunicularis* were collected from daily market and Godavari river near Paithan Maharashtra State. After bringing them to laboratory they were washed and kept in laboratory for 24 hours prior to experimentation.

To study the effect of two biogenic amines like acetylcholine and adrenalin on the oxygen consumption of crab *B. cunicularis* the following five sets of experiments were conducted.

Set I : The oxygen consumption was measured in normal crab.

Set II : 0.1 ml Acetylcholine (1/1000/w/v) was injected/crab.

Set III : 0.5 ml Acetylcholine (1/1000/w/v) was injected/crab.

Set IV : 0.1 ml Adrenalin (1/1000/w/v) was injected/crab.

Set V : 0.5 ml Adrenalin (1/1000/w/v) was injected/crab.

Before the start of experiment, pH, temperature and oxygen tension of the water were determined. The oxygen consumption was measured after 1 hour and 24 hours in all the five sets according to standard winkler's method. The weight of each crab was noted before starting of experiment to find out the oxygen consumption ml/h/1/gm body wt. Every time replication of three were run and the average values were determined.

OBSERVATION/RESULTS

The data for the amount of oxygen consumed by the experimental crabs in different sets is given in Table 1 and 2. It is observed that the injection of low dose (0.1 ml/crab) of acetylcholine caused very little increase in the oxygen consumption but the high dose (0.5 ml/crab) caused much increase in the rate of oxygen consumption in *B. cunicularis* after 1 hour and 24 hour (Table 1 and 2).

Adrenalin on the other hand caused little increase by low (0.1 ml/crab) and high (0.5 ml/crab) dose after 1 hour but it caused decrease in the rate of oxygen consumption after 24 hour. The rate is nearly equal to that of normal at both the concentration of adrenalin after 24 hours (Table 1 and 2).

TABLE 1 : EFFECT OF DIFFERENT CONCENTRATIONS OF BIOGENIC AMINES ON OXYGEN CONSUMPTION OF *B. CUNICULARIS* AFTER 1 HOUR

Temp. 32°C

O_2 tension – 3.98 ml/l

pH – 6.5

Substances injected	Concentration	Body weight (average weight in gm)	Oxygen consumption (ml/hr/gm/l) ± S.D.
Control	0	38.805	1.52 ± 0.0092
Acetylcholine	0.1 ml/crab	38.520	1.50 ± 0.0102
Acetylcholine	0.5 ml/crab	35.226	1.92 ± 0.0089
Adrenalin	0.1 ml/crab	35.465	1.600 ± 0.0072
Adrenalin	0.5 ml/crab	36.288	1.68 ± 0.0086

TABLE 2 : EFFECT OF DIFFERENT CONCENTRATIONS OF BIOGENIC AMINES ON OXYGEN CONSUMPTION OF *B. CUNICULARIS* AFTER 24 HOURS

Temp. 32°C

O_2 tension – 3.98 ml/l

pH – 6.5

Substances injected	Concentrations	Body weight (average weight in gm)	Oxygen consumption (ml/hr/gm/l) ± S.D.
Control	0	38.805	1.42 ± 0.0054
Acetylcholine	0.1 ml/crab	38.520	1.56 ± 0.0048
Acetylcholine	0.5 ml/crab	35.246	1.97 ± 0.0072
Adrenalin	0.1 ml/crab	35.415	1.35 ± 0.0039
Adrenalin	0.5 ml/crab	36.488	1.32 ± 0.004

DISCUSSION

The principle which regulate respiration in crustaceans is non neural in nature that is hormonal. (Hanumante *et. al.* 1979 b). The neurohormonal control of respiration is operational in several piokilotherms. Molluscan bearts have proved valuable for assay of acetylcholine, 5-hydroxytryptamine and noradrenalin. Neurons responsive to acetylcholine, 5-hydroxytryptamine, noradrenalin, dopamine and histamine have been detected in molluscan ganglia. The natural occurrence of acetylcholine, adrenalin and 5-hydroxytryptamine has been demonstrated in oligochaetes. (Bianchi, 1962, Laverack, 1963).

Acetylcholine is the neurohormonal transmitter at a variety of neuroeffector junctions and synoptic sites. The injection of low dose (0.1 ml/crab) acetylcholine did not cause much increase in the rate of oxygen consumption in *B. cunicularis*, but with the higher dose the rate was elevated after 1 hour and 24 hours.

The injection of low dose (0.1 ml/crab) and high dose (0.5 ml/crab) of adrenalin (epinephrine) caused little increase in the oxygen consumption after 1 hour in *B. cunicularis* but it caused decrease in oxygen consumption after 24 hours time. It can be suggested that the above increase and decline may be because of it's peripheral action.

Adrenalin when administered at higher dose caused slight elevation in the rate of oxygen consumption in *Lymnaea auricularia* (Vipradas, 1981). The elevation in oxygen consumption may be due to cardioaccelerator and/or vasodilatory influence of adrenalin (Gorbman and Bern, 1974, Laverack, 1963).

Since the endogenous presence of these biogenic amines in crustaceans ganglia has already been evidenced, regulation of respiratory metabolism may be tentatively postulated as one of the physiological roles for these neurohormones in the metabolic process of crustaceans.

SUMMARY AND CONCLUSION

1. Considering the role played by different biogenic amines in metabolic and physiological activities the study has been undertaken to find out their effects on the oxygen consumption of crab *B. cunicularis*.

2. The injection of low dose (0.1 ml/crab) of acetylcholine did not cause much increase in the rate of oxygen consumption but with the higher dose (0.5 ml/crab) the rate was elevated after 1 hr. and 24 hr. in *B. cunicularis*.

3. The injection of low (0.1 ml/crab) and high (0.5 ml/crab) doses of adrenalin caused little increase in the oxygen consumption in *B. cunicularis* after 1 hr. but it caused decrease in oxygen consumption after 24 hrs.

ACKNOWLEDGEMENTS

The author is grateful to Dr. R. Nagabhushanam, Prof. & Head, Department of Zoology, Marathwada University, Aurangabad, for providing necessary facilities. Thanks are also due to Dr. U.H. Mane, Reader in Endocrinology, Department of Zoology, Marathwada University, Aurangabad, Maharashtra.

REFERENCES

Aramant, Robert and Elofsson, Rolf (1976), Neurons in the nervous system of crustacean cell. Tiss. Res. **170** : 231-251.

Bianchi, S. (1962), Surla presence dun material fluorescent danstes du system nerveux central de quelques oligochaetes. Research histochimiques et D' histo Apwxreofeaphied - e fluorescence. Annis, Histochem. **2**: 143-145.

Cottrell, G.A. and Laverack, M.S. (1968), Annu. Rev. Pharmacol 8, 273.

Fingerman, S.W. and Fingerman, M. (1977), Circadian variation in the levels RPDH 5-HT in the eyestalks of Fiddler crab *Uca pugilator* Comp. Biochem. Physiol. 56 C, 5-8.

Gorbman, A. and Bern, M.A. (1974), A text book of comparative endocrinology, Wiley Eastern Private Ltd., New Delhi.

Laverack, M.S. (1963), Physiology of earthworms Pergamon Press Oxford London, New York, Paris.

Nagabhushanam, R., Hanumante, M.M. and Farooqui, U.M. (1976b), Effect of pharmacological and endocrinological stimulations on the respiration of marine crab *scylla serrata*. India. J. Exp. Biol. **17**(9): 964-965.

Scneider, H.P.G. and Maccann, S.M. (1969), Possible role of dopamine as transmitter to promote discharge of LH-releasing factor, Endocrinology, **85**: 121-132.

Tanc, L. (1966) In "Physiology of Mollusca" (K.M. Wilbur and C.M. Yonge eds.) Vol. 2, pp. 287-454.

Vipradas, K.C. (1981), Ph. D. Thesis. Marathwada University, Aurangabad, Chap. 5, pp. 200-210.

10
Role of Some Environmental Factors on the Fluctuation of Plankton in a Lentic Pond at Calcutta

S.K. Sarkar & P. Basu Chowdhury

INTRODUCTION

Elaborate investigations over the last four decades on the limnobiology of different types of freshwater ecosystems such as ponds, rivers and lakes reveal the interaction of various types of parameters which influence the productivity of a particular ecosystem. Earlier reports have shown that more importance has been given on limnology pertaining to large lentic and lotic water bodies by several workers including India (Ganapati, 1943; Moyle, 1946; Nygaard, 1949; Hutchinson, 1957; Phillipose, 1959; Margalef, 1964; Zafar, 1967; Michael, 1968; Patalas, 1972; Gachter *et al.*, 1974; Bharati and Kore, 1975; Kelly, 1978; Singh and Sahai, 1978; Jana, 1979; Jana *et. al.*, 1980; Khatri, 1984,1992; Dutta *et. al.*, 1982, 1983; Negi and Pant, 1983; Adoni and Saini, 1984; Madhyastha, 1986; Sharma *et al.*, 1986; Kundu and Sarkar, 1987; Birasal *et. al.*, 1989; Saha and Pandit, 1990; Nirmal Kumar, 1991; Kaushik *et al.*, 1992; Pandy *et. al.*, 1992; Srivastava, 1993; Bais *et. al.*, 1993; Nayak and Bharona, 1993; Bilgrami *et al.*, 1994; Hazarika and Dutta, 1994; Singh *et al.*, 1996). In these studies detailed investigations have taken up on various aspects of freshwater ecosystem such as (1) density and diversity of plankton population and bottom biota, (2) trophic and nutrient status, (3) temporal, spatial and seasonal distribution of plankton and bottom fauna, (4) seasonal diurnal rhythm of physico-chemical parameters, plankton and bottom fauna, (5) eutrophication evaluation by applying physico-chemical indices and plankton species, (6) seasonal fluctuation and inter-relationship between plankton and water parameters, (7) nutrient budget and primary productivity, (8) population dynamics of plankton and benthos, (9) water quality assessment and (10) taxonomic studies of plankton and bottom fauna. This is significant in the understanding of the relationship between plankton population and fish yield. In India, a very little work has been made on the seasonal changes of plankton population in freshwater ponds of smaller dimensions (Singh and Sahai, 1978; Jana, 1979; Jana *et al.*, 1980; Dutta *et al.*, 1982, 1983; Madhyastha, 1986; Pandy *et al.*, 1992; Hazarika and Dutta, 1994; Singh *et al.*, 1996) and therefore, more information is necessary on pond ecosystem because in these small water areas fish culture practices on scientific basis can easily be manipulated for maximum production. Moreover, small aquatic ecosystems are very important so far as socio-economical aspect is concerned. However, it is not possible to make a scientifically justifiable generalization on various aspects of limnology. The present study deals with the quantitative and qualitative changes of plankton population in different months of the year 1992 in a lentic (standing) fish pond situated in Calcutta, India.

MATERIALS AND METHODS

The experimental pond is perennial and situated close to College campus (22°34'N Lat. 88°30'E Long.). The pond is rectangular (125 x 20 m) and the average depth is 2.8m with a mean height of 9.1m above the sea level. The level of water of pond varied between 2.5-5.3m in different months of year. Allochthonous organic matter of the pond is considerably increased approximately all the year round due to human activities on the edge of the pond. Generally, the pond is used for fish culture of various species of major and exotic carps. The climatic condition is typical, with average air temperature of 18.6-34.7°C, 0-268mm rainfall and 54-87% relative humidity throughout the year.

Bottom sediments, surface water and plankton samples were collected between 10 a.m. and 2 p.m., once in a month from January 1992 to December 1992. Analysis of water and mud were made according to the methods described in APHA (1985). Plankton populations were collected by filtering 70 l of water through a net of bolting silk No. 21 taken from several areas of the pond, selected at random. Qualitative and quantitative estimations of plankton were made by wet weight and drop methods (APHA, 1985). The general species diversity index was calculated according to Shannon formula as follows :

$$H = -\sum\left(\frac{ni}{N}\right) \log\left(\frac{ni}{N}\right)$$

or

$$-\Sigma P_i \log P_i$$

n_i = importance value for each species
N = total importance values
P_i = importance probability for each species = n_i / N

RESULTS

Physico-chemical factors of water : The temperature of water in this pond varied between 20.3°C in January and 33°C in May (Fig. 1a). From February to August, the pH of water was alkaline (pH varied between 7.5 and 8.7) while in other months, pH values varied between 6.0 and 7.0. Total alkalinity of water in different months of the year also varied greatly (values ranged from 175 to 275 mg/1). Hardness of water was high during March through August but remained low in rest of the months. Filtrable orthophosphate concentration of water was high during March through June but remained low during July to October and again increased in rest of the months. Dissolved oxygen and total ammonia nitrogen concentrations of water fluctuated greatly in different months of the year (values ranged from 5.4 to 13.0 mg/1 for dissolved oxygen and 0.12 to 0.34 mg/1 for total ammonia nitrogen) (Fig. 1b).

Physico-chemical factors of soil : The pH of soil was low (6.0 - 7.2) during March through July except during other months when pH of soil was higher (7.0 - 8.0, Table 1). Available nitrogen and available phosphorus of soil also exhibited considerable variations in different months of the year.

Phytoplankton community : Forty six species under 28 genera were reported from the pond. Monthly variations of different species of phytoplankton are shown in Table 2. However, among different groups of phytoplankton, chlorophyceae and

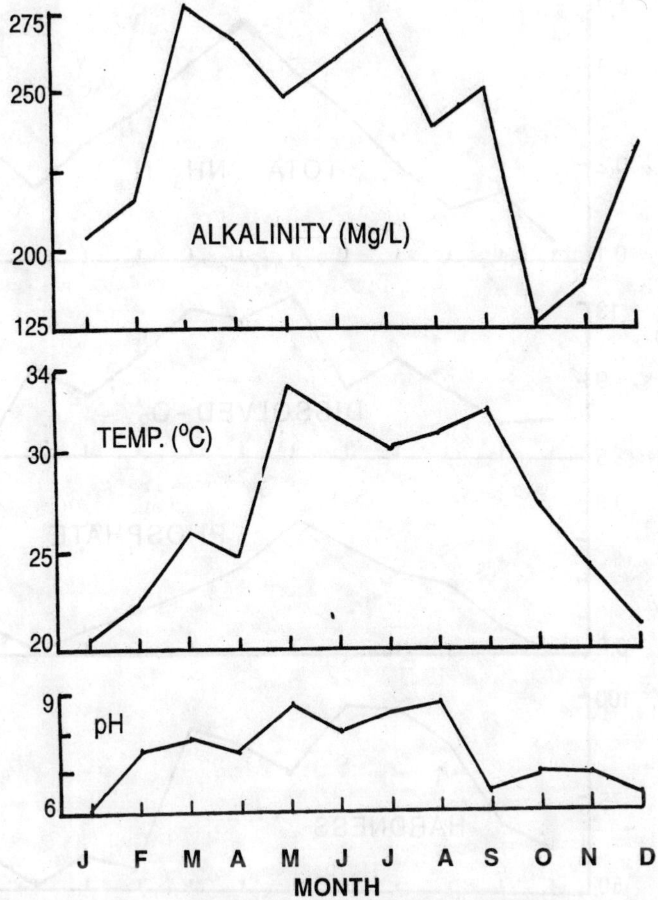

FIG. 1a : MONTHLY VARIATIONS OF PHYSICO-CHEMICAL PARAMETERS OF POND WATER, EXCEPT pH AND TEMPERATURE. ALL OTHER PARAMETERS ARE IN TERMS OF Mg/L

euglenophyceae exhibited highest and lowest concentrations, respectively and the sequence of dominance of different groups of phytoplankton was as follows : Chlorophyceae (42.6%) > Cyanophyceae (31.0%) > Bacillariophyceae (18.0%) > Euglenophyceae (9.4%). Concentration of phytoplankton (number/1) greatly differed between any two successive months. Variation of phytoplankton concentration from minimum to a maximum occurred in different months of the year in the pond was in the following order : May > April > March > December > November > January > June > July >

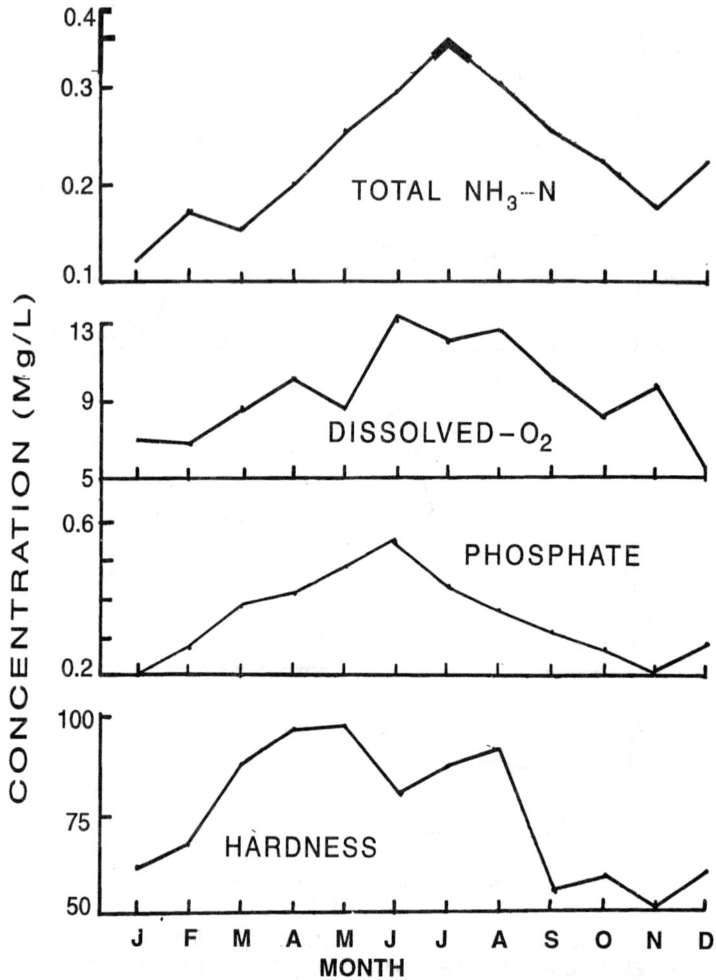

FIG. 1b : MONTHLY VARIATIONS OF PHYSICO-CHEMICAL PARAMETERS OF POND WATER. ALL PARAMETERS ARE IN TERMS OF Mg/L

February > October > August > September for Chlorophyceae, August > April and June > November > May > March > January > September > December > February > July and October for Cyanophyceae, September > June and October > February > August > December > January > November > March > July > May > April for Bacillariophyceae and February > September > October > July > January > August and December > June > March > November > April and May for Euglenophyceae.

TABLE 1 : MONTHLY VARIATIONS OF PHYSICO-CHEMICAL PARAMETERS OF BOTTOM SEDIMENTS IN POND

Parameters	Month							
	Jan.	Feb.	Mar.	Apr.	May	June	July	Aug.
pH	7.4	7.7	6.0	6.5	7.0	7.2	6.8	7.2
Available Nitrogen (kg / ha)	0.675	0.747	0.700	0.814	0.615	0.650	0.770	0.800
Available phosphorus (kg / ha)	0.475	0.575	0.534	0.500	0.485	0.520	0.620	0.645

Parameters	Sep.	Oct.	Nov.	Dec.
pH	7.5	7.5	8.0	7.8
Available Nitrogen (kg / ha)	0.850	0.765	0.875	0.800
Available phosphorus (kg / ha)	0.620	0.655	0.600	0.585

TABLE 2 : OCCURRENCE OF PHYTOPLANKTON POPULATION IN THE POND

Species	Month											
	J	F	M	A	M	J	J	A	S	O	N	D
CHLOROPHYCEAE												
Chlorella vulgaris Beijer	A	F	A	A	A	A	L	L	L	L	A	A
Chlosteridium lunula Reinsed	F	R	R	R	L	F	R	L	R	F	F	F
Clasterium setaceum Ehr	R	L	R	F	F	R	R	R	L	L	L	R
C. acerasum Ehr	L	R	F	R	F	F	F	R	L	L	L	F
Coelestrum sphericum Naeg	A	F	A	A	A	A	F	F	F	F	A	A
C. microporum Naegeli	A	F	A	A	A	A	F	F	F	F	A	A
Desmidium swartzii Agaredh	R	R	L	L	F	F	L	L	R	R	L	R
Pandorima morum Bory	L	R	F	F	F	A	R	F	F	R	L	R

(Contd.)

Species	Month											
	J	**F**	**M**	**A**	**M**	**J**	**J**	**A**	**S**	**O**	**N**	**D**
Pediastrum boryanum Menegh	R	F	R	A	F	–	A	–	F	–	L	–
Scenedesmus obliguas Turp	L	A	L	A	R	F	A	R	A	R	–	A
S. abundanus Kirchner	–	R	–	L	L	–	L	R	–	–	R	–
S. bijuga Turp	A	–	R	A	R	R	L	L	R	R	L	–
S. falcatus Chodat	L	R	F	R	F	L	F	–	–	R	L	A
Selenestrum westi Fritsch	–	R	–	A	R	A	R	A	F	A	F	R
S. gracile Reinsch	–	R	L	R	F	R	–	–	–	–	–	R
Tetradesmus cunbricus West	A	–	F	F	–	F	A	F	F	R	A	L
Tetraedron trigonum Naeg	R	L	F	A	F	L	F	R	L	L	–	R
Volvox areus Ehr	A	L	A	A	A	A	F	L	F	F	A	A
Cosmarium granatum Breb	–	A	–	F	–	R	A	–	F	–	–	–
Pediastrum tetras Ehr	A	F	A	R	–	F	A	–	F	F	–	–
P. simplex Neyen	–	–	R	–	R	A	F	R	A	F	A	A
CYANOPHYCEAE												
Anabaena affinis Lemm	F	L	A	L	A	F	R	A	–	L	A	–
A. spirodes Kleb	R	–	A	R	A	F	F	A	L	R	A	R
Aphanocapsa pulchrun Kuetz	–	–	–	R	R	–	L	R	–	–	R	R
Calothrix fusca (Kuetz) Borr et. Flah	L	L	F	F	R	–	–	R	–	F	L	F
Gleocapsa rupestris Kuetz	F	F	L	L	–	L	F	R	L	R	–	R

(Contd.)

Species	Month											
	J	F	M	A	M	J	J	A	S	O	N	D
Nostoc sphaericum Vaucher	L	F	–	F	R	–	–	F	L	L	L	L
Oscillatoria limosa Agradh	–	–	A	R	A	F	F	A	L	L	A	F
C. tenuis Agradh	R	–	A	R	A	F	F	A	L	L	A	F
O. lacustris Kleb	–	–	A	R	A	L	L	A	F	F	A	R
Spirulina gigantaea Schemidle	–	F	F	F	–	–	R	–	F	–	F	F
S. maior Kuetz	–	F	L	L	–	F	R	–	L	–	–	–
Microsystis aeruginosa Kuetz	F	F	A	L	A	L	–	A	L	L	A	F
BACILLARIOPHYCEAE												
Cyclotellla bodanica Eul	F	F	L	L	L	–	A	R	A	A	R	–
C. stelligera Clave	L	F	R	R	–	L	A	L	A	A	R	L
Cymbella affinis Kuetz	R	–	–	–	R	L	–	R	L	–	–	–
Navicula oblong Kuetz	F	F	R	L	R	R	A	R	R	A	A	R
N. bacillum Ehr	L	–	R	R	L	R	A	R	A	A	R	R
N. semilunum Grun	L	L	R	–	F	R	A	R	A	A	–	–
N. viridula Kuetz	L	F	R	–	R	L	A	L	A	A	–	–
Melosira varians Ag	L	F	–	F	–	F	A	L	A	A	R	R
EUGLENOPHYCEAE												
Euglena acus Ehr	R	A	R	L	–	–	R	–	A	A	L	L
E. polymorpha Dang	F	A	L	L	R	R	–	L	A	A	R	–
E. viridis Ehrenb	L	A	F	F	–	F	R	R	A	A	F	F
Phacus longicauda Ehr	R	A	–	F	–	–	R	L	A	A	–	R
P. caudatus Tenuis	R	A	–	–	R	F	–	L	A	A	R	L

A = Abundant, L = Less abundant, F = Frequent, R = Rare

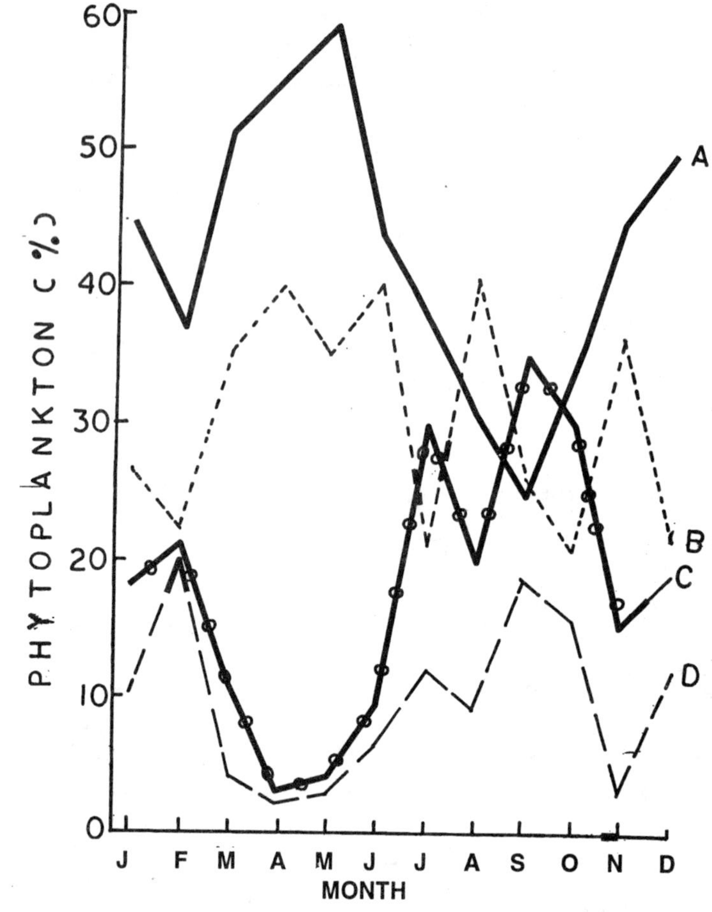

FIG. 2 : MONTHLY VARIATIONS OF DIFFERENT GROUPS OF PHYTOPLANKTON IN POND A, CHLOROPHYCEAE; B, CYANOPHYCEAE; C, BACILLARIOPHYCEASE, D. EUGLENOPHYCEAE

(a) Chlorophyceae : Fourteen genera under 21 species were observed in the pond. The genera encountered in the pond was in the following order of dominance : *Chlorella* > *Clasteriun* > *Chlosteridium* > *Coelastrum* > *Pandorina* > *Scenedesmus* > *Desidium* > *Selenestrum* > *Volvox* > *Tetradesmus* > *Tetraedron* > *Cosmarium* > *Pediastrum*. The value of Shannon index was — 1.1542 in the pond and the per cent importance of different species of Chlorophyceae versus $P_i \log_e P_i$ is plotted in Fig. 2.

It has been observed that concentration of Chlorophyceae was higher from March to June (Fig. 3) in the pond (44.0-58.5% of the total green algal population), caused due to intensive development of *Chlorella* sp. (3,250 cells/1), *Volvox* sp. (2,670 cells/1) and *Coelestrum* sp. (1,990 cells/1). Therefore, the population of chlorophyceae was gradually declined and again increased from November to January (44.6-44.9% of the total green algal population).

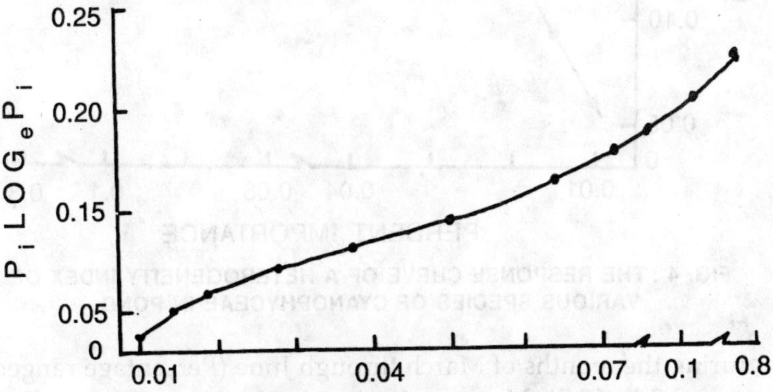

FIG. 3 : THE RESPONSE CURVE OF A HETEROGENEITY INDEX OF VARIOUS SPECIES OF CHLOROPHYCEAE IN POND

(b) Cyanophyceae : Eight genera under 12 species were observed in the pond. Different genera of Cyanophyceae encountered in the samples were in the following order of dominance : *Anabaena* > *Mycrosystis* > *Oscillatoria* > *Calothrix* > *Spirulina* > *Nostoc* > *Gleocapsa* > *Aphanocapsa*. The value of Shannon index was -0.8023 and the percent importance of different species of Cyanophyceae versus $P_i \log_e P_i$ is shown in Fig. 4.

A distinct quadrople peak was observed with main algal bloom in the months of August (40.5%) and November (37.2%) and two other peaks in March (34.7%) and May (35.0%) (Fig. 3). The development of these peaks was due to excessive blooms of *Microsystis* sp., *Anabaena* spp. and *Oscillatoria* spp. (1.8 g/1).

(c) Bacillariophyceae : In contrast to Chlorophyceae and Cyanophyceae, the concentration of Bacillariophyceae was less. However, the concentration of Bacillariophyceae was very low

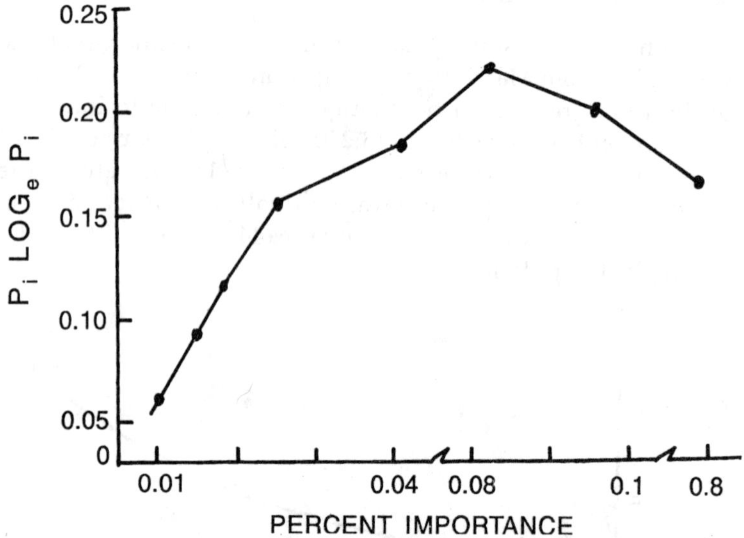

FIG. 4 : THE RESPONSE CURVE OF A HETEROGENEITY INDEX OF VARIOUS SPECIES OF CYANOPHYCEAE IN POND

during the months of March through June (Percentage ranged from 3.0 to 10.5). Maximum concentrations were observed in July, September and October (30.0, 35.0 and 30.0%, respectively). This is mainly due to excessive development of *Melosira* sp. (335 cells/1), *Navicula* spp. (450 cells/1) and *Cyclotella* spp. (290 cells/1). The genera encountered was in the following order of dominance : *Melosira* > *Navicula* > *Cyclotella* > *Cymbella*. However, only four genera under 8 species were reported from water samples.

(d) Euglenophyceae : Only two genera under 5 species have been observed in the samples and the most strongly genera encountered were in the following order of dominance : *Euglena* (160 cells/1) and *Phacus* (140 cells/1). This group of algae exhibited three peaks : one in February and two successive peaks in September and October.

Zooplankton community. Seventeen species under 13 genera were observed during the period of study in the pond. Monthly variations of different species of zooplankton are shown in Table 3. The population of zooplankton in the pond was always smaller than that of phytoplankton. Variations of zooplankton

TABLE 3 : OCCURRENCE OF ZOOPLANKTON POPULATION IN THE POND

Species	Month											
	J	F	M	A	M	J	J	A	S	O	N	D
ROTIFERA												
Brachionus angularis Gosse	L	A	A	A	A	A	L	L	R	–	–	L
B. rubens Ehr	L	A	A	A	A	A	L	–	R	L	L	R
B. calyciflorus Pallas	L	A	A	A	A	A	L	R	L	L	L	–
Keratella tropica Apstein	L	A	A	A	L	L	L	F	L	L	F	L
K. valga Ehr	L	A	L	A	L	L	L	–	–	L	R	L
Asplanchna sp.	R	–	R	–	–	–	–	–	R	–	–	–
Filinia longiseta	L	L	A	F	A	A	L	L	L	–	–	R
Rotaria sp.	L	L	A	F	A	A	L	F	F	L	R	–
Trichocera longiseta	–	R	–	R	–	R	–	–	–	–	–	–
T. similis	–	–	–	R	–	–	–	–	R	–	–	R
CLADOCERA												
Moina brachiata Jurine	A	A	L	A	L	L	L	L	L	A	L	A
Cariodaphnia cornuta Sars	R	–	–	–	–	–	R	–	R	–	–	R
Daphnia magna	A	F	A	L	A	A	L	A	L	A	L	A
Bosmina longirostris	R	–	–	–	–	–	R	–	R	–	–	R
COPEPODA												
Tropocyclops prasinus Fischer	R	L	L	A	A	A	R	A	A	F	R	R
Mesocyclops hyalinus Rehberg	R	R	–	A	A	A	A	A	A	R	–	R
Heliodiaptomus viduus Gurney	–	–	–	R	–	R	–	–	–	R	–	–
Nauplii larva	L	R	--	A	A	A	A	A	A	F	F	R

A = Abundant (201 and above), L = Less abundant (101-200),
F = Frequent (41-100), R = Rare (1-40)

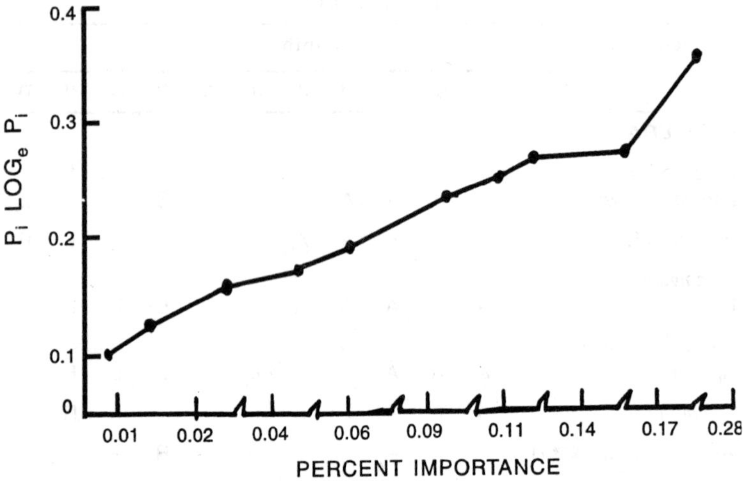

FIG. 5 : THE RESPONSE CURVE OF A HETEROGENEITY INDEX OF VARIOUS SPECIES OF ZOOPLANKTON IN POND

concentration in different months of the year in the pond was in the following order of dominance : May > June > March > February > April > September > August and January > November and December > July > October for Rotifera, June > August > January > November > December > October > February and July > September > May > April > March for Cladocera and April > May > August > September > June > July > October > March > February and November > January > December for Copepoda. The value of Shannon index for different species of zooplankton was -2.0521 and the response curve of heterogeneity index for zooplankton species is shown in Fig. 5.

(a) Rotifera : Six genera under 10 species were observed. Of them, *Brachious* spp., *Keratella* spp., *Filinia* sp. and *Rotatoria* sp. were most common. They occurred in large numbers in the months of February-June (202-225 units/1) and thereafter, their numbers declined in subsequent months (110-140 units/1) (Fig. 6). *Asplanchna* sp. and *Trichocera* spp. were sporadic in distribution, found only in three samples and their numbers were very less (10 units/1).

(b) Cladocera : Only four genera under 4 species were encountered in the pond. Among different species only *Moina* sp. and *Daphnia* sp. were common and observed all the year

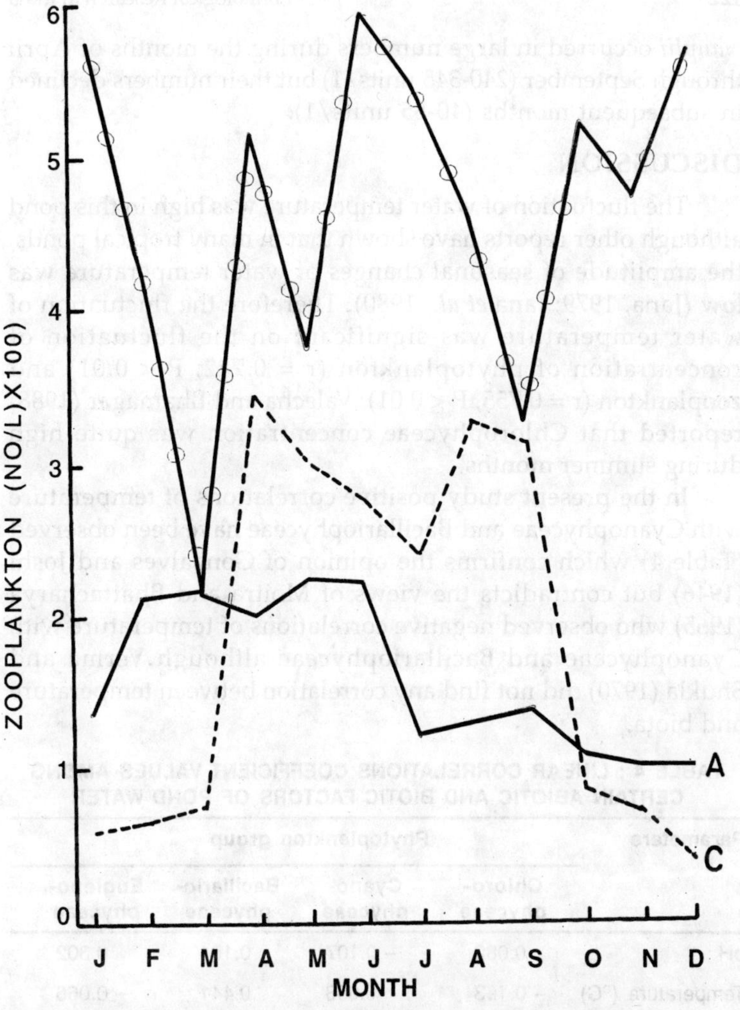

FIG. 6 : MONTHLY VARIATIONS OF DIFFERENT GROUPS OF ZOOPLANKTON IN POND. A. ROTIFERA; B. CLADOCERA; C. COPEPODA

round (200-590 units/1). *Ceriodaphnia* sp. and *Bosmina* sp. were rarely observed in the pond and their concentrations being 12 and 7 units/1 respectively, in July, September, December and January.

(*c*) *Copepoda* : Only three genera under 3 species were observed in the pond. *Tropocyclops* sp., *Mesocyclops* sp. and

Nauplii occurred in large numbers during the months of April through September (240-345 units/1) but their numbers declined in subsequent months (40-85 units/1).

DISCUSSION

The fluctuation of water temperature was high in this pond although other reports have shown that in many tropical ponds, the amplitude of seasonal changes of water temperature was low (Jana, 1979; Jana *et al.*, 1980). Therefore the fluctuation of water temperature was significant on the fluctuation of concentration of phytoplankton ($r = 0.782$, $P < 0.01$) and zooplankton ($r = 0.755$, $P < 0.01$). Valecha and Bhatnagar (1988) reported that Chlorophyceae concentration was quite high during summer months.

In the present study positive correlations of temperature with Cyanophyceae and Bacillariophyceae have been observed (Table 4) which confirms the opinion of Gonzalves and Joshi (1946) but contradicts the views of Moitra and Bhattacharya (1965) who observed negative correlations of temperature with Cyanophyceae and Bacillariophyceae although Verma and Shukla (1970) did not find any correlation between temperature and biota.

TABLE 4 : LINEAR CORRELATIONS COEFFICIENT VALUES AMONG CERTAIN ABIOTIC AND BIOTIC FACTORS OF POND WATER

Parameters	Phytoplankton group			
	Chloro-phyceae	Cyano-phyceae	Baciliario-phyceae	Eugleno-phyceae
pH	0.082	– 0.107	0.104	0.302
Temperature (°C)	– 0.163	0.058	0.441	– 0.066
Total alkalinity (mg/l)	– 0.301	0.053	0.307	0.401
Hardness (mg/l)	0.060	0.596	– 0.227	– 0.019
Orthophosphate (mg/l)	– 0.397	– 0.110	– 0.162	– 0.024
Dissolved oxygen (mg/l)	0.483	0.492	0.477	0.453
Total ammonia nitrogen (mg/l)	0.107	0.121	0.150	0.101

Although pH is very important for phytoplankton growth, pH 5.0-8.5 is best for phytoplankton growth (Robert, 1974) but harmful at pH more than 8.8 (Ellis, 1937). In the present study positive correlation of pH with Chlorophyceae, Bacillariophyceae and Euglenophyceae and negative correlation with Cyanophyceae was observed. Positive correlations between pH and phytoplankton groups have been reported by Laxminarayana (1965).

A direct relationship among dissolved oxygen (DO) and different groups of algae have been observed. Verma and Mohanty (1995) observed positive correlations between DO and Bacillariophyceae and Euglenophyceae and negative correlations between DO and other groups of algae.

It has been established that DO concentration was dependent upon the phytoplankton concentration. The regression line was linear (Fig. 7). A distinct positive correlation between total ammonia nitrogen, phosphate and different

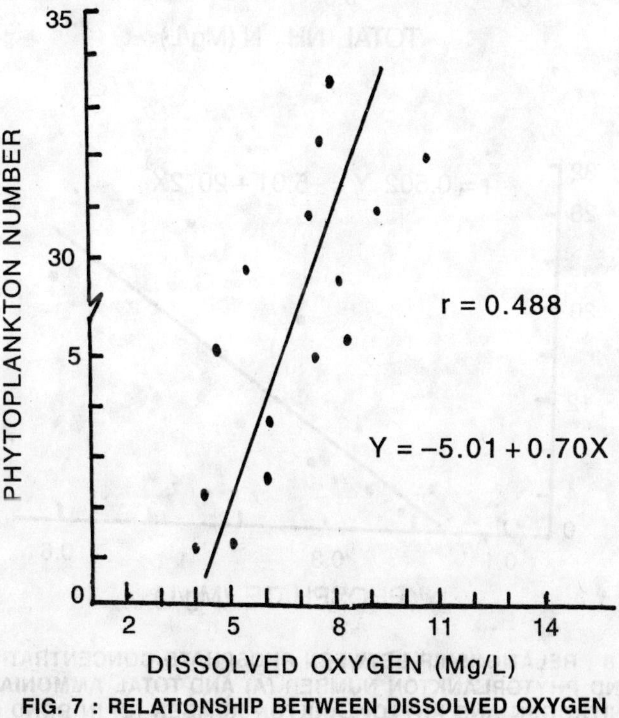

FIG. 7 : RELATIONSHIP BETWEEN DISSOLVED OXYGEN CONCENTRATION AND PHYTOPLANKTON NUMBER IN POND

groups of algae were observed in the pond and their general relationships are shown in Fig. 8. Although inorganic soluble phosphate is assimilated by algal blooms (Kramer *et al.*, 1979). Schandler (1979) did not find any significant correlation between phosphorus and plankton.

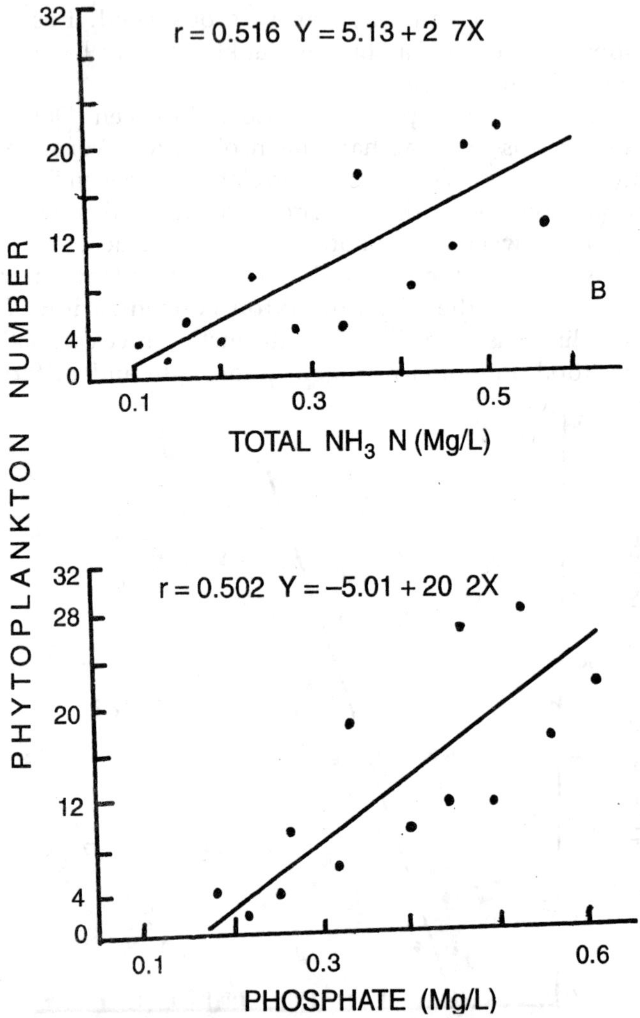

FIG. 8 : RELATIONSHIP BETWEEN PHOSPHATE CONCENTRATION AND PHYTOPLANKTON NUMBER (A) AND TOTAL AMMONIA NITROGEN AND PHYTOPLANKTON NUMBER (B) IN POND

Since the monthly fluctuations of phytoplankton was the inverse functions of total alkalinity of water ($r = 0.51, P < 0.05$), it is reasonable to conclude that the concentration of total alkalinity was an alternative source of carbon for photosynthesis. Other workers (Sarkar and Rai, 1964; Jana *et al.*, 1980) have observed similar results in ponds. Jhingran (1988) opined that the natural water bodies in the tropics exhibited wide range of variations in the total alkalinity.

Rotifers exhibited superiority over cladocera and copepoda. Similar observations have also been made in other freshwater bodies (Michael, 1968; Singh and Sahai, 1978; Sarwar and Parveen, 1996). The sequence of dominance of different groups of zooplankton was : Cladocera > Copepoda > Rotifera. But so far as the number of different species of zooplankton are concerned it was observed that the number of Rotifer species was highest which was followed by Cladocera and Copepoda. However, the abundance of rotifers and cladocera might be due to the dependence on phytoplankton. Increase in the concentrations of rotifera and copepoda during summer months were associated with the increase in phytoplankton populations. Watson and Wilson (1978) observed higher concentration of zooplankton at the time when phytoplankton concentration increased in several ponds. Sarwar and Parveen (1996) opined that higher zooplankton number was recorded during summer in some tropical lakes. Decrease in concentrations of Rotifera and Copepoda may be related to low water temperature (Sarwar and Parveen, 1996).

It is in general agreement that zooplankton and phytoplankton are closely related to each other in aquatic ecosystem and therefore, it is not observed in pond ecosystem where zooplankters consume phyoplankton and nanoplankton. Ponds when stocked with fish, often caused a drastic reduction in the concentration of plankton biomass. Although no correlation was made between grazing of fish and production of nanoplankton, insignificant correlation ($r = 0.507, P > 0.05$) between zoo- and phytoplankton was the result whether the relationships were based on the monthly observed data. Monthly fluctuations of different groups of zooplankton do not point a definite relationship among themselves (Fig. 6).

The multiple correlations involving the fluctuations of phytoplankton number and the concentration changes of total

TABLE 5 : MULTIPLE CORRELATION COEFFICIENTS
Y = PHYTOPLANKTON NUMBER/1 (PN), X_1 = TOTAL AMMONIA NITROGEN, X'_1 = TOTAL ALKALINITY, X_2 = PHOSPHATE, X_3 = DISSOLVED OXYGEN

Relationships	Regression equation	$R_{Y1.23}$	F	Final regression equation
PN Vs. Concentration of X_1, X_2 and X_3	$Y = 48.01 (a) - 0.150 X_1 - 140.2 X_2 - 7.72 X_3$	0.812	5.20*	$Y = -5.84 (a) + 33.78 X_2$
PN vs. rate of concentration changes of X_1, X_2 and X_3	$Y = 51.37 (a) - 0.274 X_1 + 54.2 X_2 - 4.57 X_3$	0.867	7.11*	$Y = 51.37 (a) - 0.274 X_1 + 54.2 X_2 - 4.57 X_3$
PN vs. concentration of X'_1, X_2 and X_3	$Y = -0.866 (a) - 0.191 X_1 + 5.50 X_2 + 0.180 X_3$	0.906	21.11***	$Y = -0.507 (a) + 2.412 X_2$

*** P < 0.05, *** P < 0.001**

TABLE 6 : MULTIPLE CORRELATION COEFFICIENTS
Y = ZOOPLANKTON NUMBER/1 (ZN), X_1 = TOTAL AMMONIA NITROGEN, X_2 = PHOSPHATE, X'_1 = TOTAL ALKALINITY, X'_2 = DISSOLVED OXYGEN, X'_3 = TEMPERATURE; X_3 = pH

Relationships	Regression Equation	$R_{Y1.23}$	F
ZN vs. concentration of X'_2 and X'_3	$Y = 683 (a) + 2.27 X'_2 - 35.52 X'_3$	0.774	40.44 (S)
ZN vs. concentration of X'_1 and X'_2	$Y = 220.9 (a) + 0.297 X'_1 - 0.767 X'_2$	0.677	19.03 (S)
ZN vs. concentration of X_1 and X_2	$Y = 203.1 (a) + 118.7 X_1 - 205 X_2$	0.790	4.50 (NS)
ZN vs. concentration of X'_2 and X_3	$Y = 201.7 (a) + 0.172 X'_2 - 0.817 X_3$	0.682	27.25 (S)
ZN vs. concentration of X'_2 and X_2	$Y = -509 (a) + 88.33 X'_2 - 10.90 X_2$	0.811	4.53 (NS)

S : Significant at 5% level; NS : Non significant.

alkalinity (TA), total ammonia nitrogen (TAN), orthophosphate (OP) and dissolved oxygen (DO) were significant (Table 5, $P < 0.05$). Moreover, multiple correlations of Y and the rate of concentration changes of TAN, OP and DO were also significant ($P < 0.05$). Therefore, the role of OP was maximum on the fluctuation of phytoplankton in the pond because partial regression coefficient did not justify the inclusion of TA, TAN and DO in the final regression equations. On the other hand, the rate of changes of concentration of TAN, OP and DO are important as these values were observed in the final regression equation.

The multiple correlation involving the fluctuations of zooplankton number and the concentration changes of dissolved oxygen, temperature, total alkalinity and pH were significant (Table 6). Obviously, the role of these factors are important on the dynamics of zooplankton concentration in the pond.

SUMMARY

1. Seasonal distribution of the population structure of phytoplankton and zooplankton was studied in connection with the physico-chemical parameters in a lentic pond over a period of one year.
2. The concentration of different species of phyto- and zooplankton varied. A distinct seasonal response was observed by phytoplankton population.
3. The fluctuation of water temperature was significant on the fluctuation of plankton concentration. A positive correlation of temperature with Cyanophyceae and Bacillariophyceae was observed. A positive correlation of pH with Chlorophyceae, Bacillariophyceae and Euglenophyceae and negative correlation with Cyanophyceae. Relationship between dissolved oxygen and different groups of algae was direct.
4. A positive correlationship between total ammonia nitrogen, orthophosphate and different groups of algae were observed in the pond. Monthly fluctuations of phytoplankton was the inverse function of total alkalinity of water.
5. The general diversity of Chlorophyceae shows a value of - 1.1542 in the pond. Similarly, the Shanon index for species of Cyanophyceae and Zooplankton varied in the pond.
6. The role of phosphate was found to be maximum on the number of phytoplankton as the total ammonia nitrogen and dissolved oxygen were not retained in the final

regression equation. On the other hand, the concentration of total ammonia nitrogen, phosphate and dissolved oxygen was greatly influenced on the number of phytoplankton as these three factors were retained in the final regression equation.

7. The multiple correlation involving the fluctuations of zooplankton number and the concentration changes of dissolved oxygen, temperature, total alkalinity, total ammonia nitrogen, phosphate, and pH were significant.

ACKNOWLEDGEMENTS

Authors are thankful to the Principal of NND College for laboratory facilities and to Shri Ashok Banerjee, Knushna Sahu and Amalendu Chakraborty for helping the authors during the period of study. Thanks are also due to Dr. S.K. De, Department of Mathematics, for statistical analysis of experimental data.

REFERENCES

Adoni, A.A. and Saini, R.S. (1984). Eutrophication and production studies of some lentic ecosystem around Sagar, M.P. Proc. 2nd Ann. Workshop on MAB Project, New Delhi, 23-25 March, 163-167.

APHA (American Public Health Association) (1985). Standard Methods for the Examination of water and wastewater. Washington, DC.

Bais, V.S., Yatheesh, S. and Agarwal, N.C. (1993). Evaluation of the trophic status of a lake. Proceedings of the Academy of Environmental Biology, **2** : 125-131.

Bharati, S.G. and Kora, A.P. (1975). Limnological studies in ponds and lakes. Karnataka University Journal of Science, **20** : 157-167.

Bilgrami, K.S., Sheo, K. and Kumar, S. (1994). Dial variations in abiotic and biotic factors of the river Ganga at Bhagalpur. Environment and Ecology, **12** : 374-378.

Birasal, N. R., Nadkrani, V.M. and Gouder, B. Y. M. (1989). Ecology of zooplankton of a reservoir of Kali River. Journal of Freshwater Biology, **1** : 15-23.

Dutta, N. C., Banerjee, B.R. and Dar, N. K. (1982). Diurnal rhythm of physico-chemical properties and zooplankton in a freshwater pond in Calcutta. Indian Journal of Physiological and Natural Sciences, **2** : 22-27.

Dutta, N. C., Banerjee, B.K. and Banerjee, S. B. (1983). Relationship between phytoplankton and primary productivity of a freshwater pond in Calcutta. Environment and Ecology, **1** : 135-137.

Ellis, M. M. (1937). Detection and measurement of stream pollution. Bulletin of Bureau of Fisheries, Washington. **48**: 172-193.

Gachter, R., Vollenweider, R. A. and Glooschenko, W. A. (1974). Seasonal variations of temperature and nutrient in the waters of lake Ontario and Eri. Journal of Fisheries Research Board of Canada, **31** : 275-290.

Ganapati, S. V. (1943). An ecological study of a pond containing zooplankton. Proceedings of Indian Academy of Sciences, **17** : 41-58.

Gonzalves, E. A. and Joshi, B. B. (1946). Freshwater algae near Bombay. I. Seasonal succession of the algae. Journal of Bombay Natural History of Society, **46**: 154-176.

Hazarika, A. K. and Dutta, A.C. (1994). Limnological studies of two ponds of Assam. Environment and Ecology, **12** : 26-29.

Hutchinson, G. E. (1957). A treatise on limnology. II. Introduction to lake biology and limnoplankton. John Wiley Company, New York.

Jana, B. B. (1979). Temporal Plankton succession and ecology of a tank. International of Reveneger Hydrobiology, **64** : 661-671.

Jana, B. B., De, U.K. and Das, R. N. (1980). Environmental factors affecting the seasonal changes of plankton on fish ponds in India. Schweiz of Zoological **42** : 225-246.

Jhingran, V. G. (1988). Fish and Fisheries of India. Hindustan Publishing Corporation, New Delhi, India.

Kaushik, S., Agarkar, M. S. and Saxena, D. N. (1992). Distribution of phytoplankton in river water in Chambal area, M.P. Bionature, **12**: 1-7.

Kelly, M.N., Fitzpatria, L.C. and Pearson, W.B. (1978). Dynamics, primary productivity and community metabolism in Texas pond. Hydrobiologia, **50** : 177-189.

Khatri, T. C. (1984). Seasonal variation in primary production in relation to limnological features in Lakhotia lake. Proceedings of Indian Academy of Sciences, **93** : 697-702.

Khatri, T. C. (1992). Seasonal distribution of zooplankton in Lakhotia lake. Environment and Ecology, **10** : 317-322.

Kramer, J. R., Herbes, S. E. and Allen, H. E. (1979). Phosphorus: Analysis of biomass and sediment. In "Nutrients in Natural Waters" (H. E. Allen and J. R. Kramer Eds.) pp. 52-100. Wiley Interscience Publications, New York.

Kundu, S. K. and Sarkar, S. K. (1987). Studies on zooplankton with reference to copepods in river waters. Proceedings of National Seminar on Estuary Management, Trivandam, 4-5 June, 323-326.

Laxminarayana, J. S. (1965). On the phytoplankton of the River Ganges, Varanasi, India. Hydrobiologia, **25** : 119-137.

Madhyastha, M. M. (1986). Diurnal variations in the hydrobiological conditions of a tropical pond in Mangalore. Bulletin of Environmental Sciences, **3** : 46-48.

Margalef, R. (1964). Correspondence between the classical types of lake and structural and dynamic properties of their populations. International Verein of Theoretical Limnology, **15** : 169-175.

Michael, R. G. (1968). Studies on zooplankton of a tropical fish pond. Hydrobiologia, **32** : 47-68.

Moitra, S. K. and Bhattacharya, B. K. (1965). Hydrobiological factors affecting plankton production in a freshwater pond, West Bengal, India. Icthyalgia, **4** : 8-12.

Moyle, J.B. (1946). Some indices of lake productivity. Transactions of The American Fisheries Society, **76** : 322-324.

Nayak, T.R. and Bharona, T.R. (1993). Plankton as indicators of eutrophication in shallow water lakes of Panna, M.P. Proceedings of Academy of Environmental Biology, **2** : 69-75.

Negi, V. and Pant, M.C. (1983). Analysis of zooplankton of Kumaun lake. Tropical Ecology, **24** : 271-282.

Nirmal Kumar, J. I. (1991). Primary production of phytoplankton of a tank, Gujarat. Indian Journal of Botanical Society, **70**: 427-428.

Nygaard, G. (1949). Studies of some ponds and lakes. II. The quotient hypothesis and some new or little known phytoplankton organism. Danska Vidersk Selsk, **7** : 1-293.

Patalas, K. (1972). Crustacean plankton and the eutrophication of St. Lawrence lakes. Journal of Fisheries Board of Canada, **29** : 1451-1462.

Phillipose, M. T. (1959). Freshwater phytoplankton of inland fisheries. Proceedings of Symposium of Algalogy, ICAR, 272-291.

Pandy, B. C., Singh, B. K. and Rao, P. N. (1992). Species diversity of plankton in fish pond at Pune, Maharashtra. Proceedings of Academy of Environmental Biology, **1**: 167-172

Robert, D. S., Robert, W. H. and Evereft, L. G. (1974). Phytoplankton distribution and water quality indices for lake Lead (Colorado River). Phycology, **10** : 323 : 333.

Saha, L. C. and Pandit, B. (1990). Dynamics of primary productivity between lentic and lotic systems in relation to abiotic factors. Indian Journal of Botanical Society, **69** : 213-217.

Sarkar, H. L. and Rai, H. (1964). Limnology of Surajkund, India: Physico-chemical conditions and plankton. Transactions of American Microscopic Society, **83** : 260-266.

Sarwar, S. G. and Parveen. A. (1996). Community structure and population density of zooplankton in two lakes of Srinagar, Kashmir, India. Pollution Research, **15** : 35-38.

Schandler, D. W. (1979). Evolution of phosphorus limitation in lakes. Science, **195** : 260-262.

Sharma, J. P., Gupta, V. K. and Srivastava, J. B. (1986). Variations in the primary production and physico-chemical properties of a eutrophic pond. Bulletin of Environmental Science, **3** : 31-33.

Singh, M., Sinha, B.K. and Sharma, G. (1996). Diurnal migration of zooplankton in a fish pond of Patna, Bihar. Geobios, **22**: 130-134.

Singh, S.B: and Sahai, R. (1978). Fluctuation of zooplankton in relation to physico-chemical factors of a pond. Geobios, **5**: 228-230.

Srivastava, V. S. (1993). Physico-chemical studies of river Jharali, Nandunbar. Indian Journal of Environmental Protection, **13**: 109-112.

Valecha, V. and Bhatnagar, G. P. (1988). Changes of phytoplankton in relation to physico-chemical factors in lakes of Bhopal. Geobios, **15** : 170-173.

Verma, J. P. and Mohanty, R. C. (1995). Phytoplankton of Malyanta pond of Laxmisagar and its correlation with certain physico-chemical parameters. Pollution Research, **14** : 243-252.

Verma, S. and Shukla, G.R. (1970). Physico-chemical condition of a tank, Muzaffarnagar, U.P. in relation to biological productivity. Environmental Health, **12** : 110-128.

Watson, N. H. F. and Wilson, J. B. (1978). Crustacean zooplankton of lake Superior. Journal of Great Lakes Research, **4** : 481-496.

Zafar, A. R. (1967). On the ecology of aigae in fish ponds of Hyderabad, India. III - The periodicity. Hydrobiologia, **30** : 96-112.

11

River Pollution in India : An Overview

S.R. Mishra

INTRODUCTION

The high rate of increase in human population and rapid pace of industrialization have created an acute problem of disposal of waste products and the domestic wastes, municipal sewage and industrial effluents by and large are being indiscriminately discharged into the rivers and the adjoining fields without any treatment or proper treatment. This has led to pollution hazards not only to the aquatic life but also to the human life. Sewage and domestic wastes include all types of refuges such as kitchen refuge, bathroom washings, detergents, night soil etc. Some of them are quite harmful and affect the quality of water and brings about water pollution. In the present article, an overview of river pollution in India has been presented.

POLLUTIONAL STATUS OF DIFFERENT INDIAN RIVERS

Chacko and Sreenivasan (1955) studied hydrobiological characteristics of three major rivers in Madras. Banerjea and Motwani (1960) took up the study of pollution in Suvaon stream by the sugar factory effluents at Balrampur in Uttar Pradesh. Hydrobiology and fisheries of the Ennore river near Madras were studied by Chacko and Rajagopal (1962). Observations on the pollution of Yamuna river at Okhla water works, Delhi were taken by Balani and Sarkar (1965) and it was suggested that the water supplies should be regularly monitored for its free ammonia content which may be useful in fixing the water quality criteria for drinking water supply. George *et al.* (1966)

have pointed out that the wastes of various industries discharged into the river Kali contain a heavy load of organic matter. The effect of industrial wastes and sewage upon the chemical and biological composition and fisheries of the river Ganga at Kanpur was studied by Ray and David (1966). The pollution studies on the river Ganges near Kanpur were also conducted by Saxena *et al.* (1966) and they found that river water deteriorated after receiving industrial waste and sewage. Sreenivasan and Raj (1967) studied effect of certain wastes on the water quality and fisheries of rivers Cauvery and Bhavani. Physico-chemical complexes of the river Moosi in Hyderabad (India) with special reference to water pollution were noted by Venkateswarlu (1969). A study of the trophic status of Gomti river at Lucknow has been conducted by Arora *et al.* (1973). Rajagolan *et al.* (1973) studied pollution of river Subarnarekha at Ranchi. In the river Kali near Mansurpur, Verma and Dalela (1975) noted that high load of organic matter present in the wastes influenced the oxygen budget of the river greatly and changed the water and bottom quality of the river adversely. Vass *et al.* (1977) studied hydrobiological features of river Jhelum in Kashmir. Prakash *et al.* (1978) have carried out ecological study of river Jamuna at Agra while Verma *et al.* (1978) have studied the physico-chemical and biological characteristics of Kadarabad drain in Uttar Pradesh. Dakshini and Soni (1979) have determined the water quality of sewage drains entering into river Yamuna at Delhi. Verma *et al.* (1980) observed the pollution of Hindon river in relation to fish and fisheries while Badola and Singh (1981) studied the hydrobiological characteristics of the river Alaknanda of Garhwal Himalaya. The water quality of sewage drains entering into Yamuna at Agra was assessed and it was found that these drains discharged high amount of pollutants in the river causing deterioration of the quality of river water (Sharma *et al.*, 1981). Agarwal (1982) had studied the nutrient level of Chambal river at Kota. Mitra (1982) studied chemical characteristics of surface water at selected stations on the rivers Godavari, Krishna and Tungabhadra. The water quality of sewage drains entering into Chambal river at Kota had high amounts of pollutants and required suitable treatment prior to its discharge in the river to restore the existing level of organic matter and deterioration of the quality of Chambal river water and to avoid various health hazards (Agarwal, 1983). Chandra and Krishna (1983) studied the impact of tannery waste disposal on the quality of the river Ganges at Kanpur. They have pointed out that quality of water is an immediate hazard for public health. The water quality profile of Vaigai river in Madurai has been determined by Mahadevan and Krishnaswamy (1983). Patil and Patil (1983) have studied the water quality of Ulhas river in Bombay. Sikander and Tripathi (1983) observed the physico-chemical characteristics of Ganga river water at Varanasi. A short term study of pollution status of river Ganga at Kanpur region suggested that the industrial wastes and municipal sewage contribute towards the pollution of the river (Chattopadhya *et al.*, 1984). The water of Godavari river at Nanded was not found suitable for drinking due to the pouring of sewage and industrial waste (Jagdale *et al.*, 1984). Raina *et al.*, (1984) have found that water in Jhelum though is not polluted at present requires a continuous monitoring of the river for various pollution parameters to avoid risk of pollution and to create increased awareness amongst people so as not to turn the river into open sewer in view of rapid urbanization, industrialization and increased traffic influx. Potamological studies on lotic environment of river Bhagirathi in Garhwal Himalaya have been carried out by Sharma (1984) while Somashekar (1984) studied the water pollution of the river Cauvery in Karnataka which is being polluted by industrial wastes. Pollution and saprobic status of Eastern Kalinadi was studied by Verma *et al.*, (1984). Ajmal *et al.* (1985) have also studied the physico-chemical aspects of pollution in this river and observed that the same is highly polluted during winter and summer seasons. The environmental impact assessment of the river Ganges at Varanasi suggested that the pouring of untreated city sewage and industrial effluent in to the river made it polluted (Chaudhary and Ojha, 1985). Studies on the ecological impact of a newsprint factory wastes on Moovattupuzha river ecosystem has been carried out by Geevarghese and Chandrasekharan (1985). Nema and Lal (1985) have carried out pollutional studies of Sone river in Madhya Pradesh and concluded that this river is polluted in between Chachai to Diyapipal. Sangu and Sharma (1985) studied pollution in

Yamuna river at Agra and concluded that the river water has deteriorated over the years. A study of pollution in the Subarnarekha river due to aluminium factory discharge at Musi in Bihar was conducted by Mishra *et al.* (1986). Nautiyal *et al.* (1986) studied the water quality and assessed potability of Alaknanda river water at Srinagar Garhwal. They concluded that the town sewage is the major source of pollution in the river. The water characteristics of river Tungabhadra at Kurnool and Alampur suggested that the river is subjected to paper mill effluent pollution (Ramesh *et al.*, 1986). Rao and Rao (1986) have observed the pollution in Kshipra river at Ujjain due to pilgrim bathers during the Kumbha fair. The impact of distillery effluents on the water quality and ecosystem of river Reh in Doon valley has been assessed by Shanker *et al.* (1986). Dora and Roy (1987) investigated the water quality of Subarnarekha river in Bihar and found that the river is not yet grossly polluted. Distributory pattern of dissolved oxygen at the selected station in Cooum river at Madras was found to affect the aquatic fauna (Jebanesan *et al.*, 1987). Lakshmanrao and Narayanarao (1987) studied pollution of selected rivers and concluded that there are only few success stones of preserving the quality of streams in India. Mishra and Mishra (1987) studied chromate pollution in the river Jhoomar at Ranchi and discussed the hazardous effect of chromium as chromate. Ajmal and Razi-ud-Din (1988) studied pollution of Hindon river and Kalinadi in Uttar Pradesh while the hydrological profile of Hooghly sector (Ballygunge to Bandel) of river Ganga was studied by Datta *et al.* (1988). The effect of sewage drain water on the physico-chemical and bacteriological characteristics of river Ganga at Patna was studied by Ghose and Sharma (1988). They have noted that the raw sewage and sullage causes an serious pollution problem in the river. Water pollution in Niva river in Andhra Pradesh has been studied by Nandkumar *et al.* (1988). Pollution of Narmada river and measures of its control at Hoshangabad have been studied by Palharya and Malviya (1988). The seasonal rhythm in the physico-chemical properties of Nana Kosi river in Kumaun region of Uttar Pradesh has been observed by Pande *et al.* (1988) and it was suggested that siltation process should be checked for environmental equilibrium. Shah (1988) discussed the physico-chemical aspects of pollution of river

Jhelum at Kashmir and concluded that the water quality in general has deteriorated from year to year basis. Sunder (1988) has also monitored the water quality in a stretch of this river at Srinagar. Chavadi and Nalawadi (1989) have used hydro-environment indicators in the environmental impact assessment of Kali river system in Karnataka. Pollution studies on river Subernarekha around industrial belt of Ranchi have been carried out by Singh and Singh (1990) and it has been found that the urbanization and rapid industrialization of this area have deteriorated the water quality and are main causes of pollution in this river. Mishra (1990) studied the hydro-biological characteristics of Morar river in relation to plankton and productivity and noticed that river is polluted by domestic sewage, municipal sewage and other discharges along its course in the city Gwalior. The distribution of algal flora in polluted regions of Karwan river at Agra has studied by Mittal and Sengar (1991) and observed that several species showed interesting pattern of succession in relation to pollution load. Unni *et al*. (1992) have noted physico-chemical and biological characteristics of river Narmada from Amarkantak to Jabalpur. BOD and COD values indicates that the river is not under heavy pollution stress. Impact of sewage and industrial pollution on physico-chemical characteristics of water in river Betwa at Vidisha, Madhya Pradesh has observed by Mishra (1996). Singh and Mahaver (1997) have studied the role of heavy metals in riverine pollution and found that the pollutional load of river Ganga and river Gomti were maximum in respect of zink, arsenic and chromium while the physico-chemical characteristics of river Ganga near Patna were observed by Singh (1997). He noticed that most of the stretches of river are the discharge sites of the domestic sewage. As a result of which the river quality are being deteriorated day by day.

CONCLUSION

On the basis of above survey it can be concluded that major Indian rivers are facing the danger of pollution. The pollutants which are contributing to pollution are industrial effluents from different industries such as sugar, textiles, pulp, paper, tanneries, distilleries and untreated municipal domestic sewage etc. These polluting agent are continuously deteriorating the water quality of different rivers and aquatic life has also been affected. It is, therefore, very necessary to obtain accurate and timely information regarding the quality of any river system. An quick need for systematic monitoring of river pollution in order to formulate sound public policy and implement river water improvement programme.

REFERENCES

Agarwal, S.K. (1982), Eutrophication of Chambal river at Kota. Acta. Ecol., **4**: 5-9.

Agarwal, S.K. (1983), Water quality of sewage drains entering Chambal river at Kota. Acta. Ecol., *5*: 26-29.

Ajmal, M., Razi-Uddin and Khan, A.U. (1985), Physico-chemical aspects of pollution in Kali Nadi. IAWPC Tech. Annual, *XII* : 106-114.

Ajmal, M. and Razi-Uddin (1988), Studies on the pollution of Hindon river and Kali nadi (India). In "Ecology and Pollution of Indian rivers" (R.K. Trivedy ed.) pp. 87-111, Ashish Publishing House, New Delhi, India.

Arora, H.C., Chattopadhyaya, S. N., Sharma, V.P, Routh, T. and Elyas, S.I. (1973), A short term study on the eutrophication of Gomti river in Lucknow region. Proc. Symp. Environ. Poll., pp. 44-58.

Badola, S.P. and Singh, H.R. (1981), Hydrobiology of the river Alaknanda of the Garhwal Himalaya. Ind. J. Ecol., *8* : 269-276.

Balani, M.C. and Sarkar, H.L. (1965), Some observations on pollution of Yamuna river at Okhla water works intake, Delhi. Environ. Hlth., **VII:** 84-86.

Banerjea, S. and Motwani, M.P. (1960), Some observations on pollution of the Suvaon stream by the effluents of a sugar factory, Balrampur (U.P.). Ind. J. Fish., ***VII*** : 107-128.

Chacko, P.I. and Rajagopal, A. (1962), Hydrobiology and fisheries of the Ennore river near Madras from April 1960 to March 1961. Madras J. Fisher., ***1*** : 102-104.

Chacko, P.I. and Sreenivasan, R. (1955), Observations on hydrobiology of three major rivers of Madras state, South India. Contr. F.W. Biol. Stn. Madras, ***13*** : 1-16.

Chandra, S. and Krishna, G. (1983), Effect of tannery waste

disposal on the quality of the river Ganges at Kanpur. Poll. Res., **2** : 63-64.

Chattopadhya, S.N., Routh, T., Sharma, V.P., Arora, H.C. and Gupta, R.K. (1984), A short term study on the pollutional status of river Ganga in Kanpur region. Ind. J. Environ. Hlth., ***20*** : 224-255.

Chaudhary, U.K. and Ojha, C.S.P. (1985), Environmental impact assessment of the river Ganges at Varanasi. Proc. Int. Sem. Environ. Imp. Assess. Wat. Res. Proj., Univ. of Roorkee, 12-14 Dec., pp. 760-771.

Chavadi, V.C. and Nalawadi, A.R. (1989), Hydro-environmental indicators : A review and evaluation of their use in the assessment of environmental impacts on Kali river system. Manag. Aqu. Ecol., 189-205.

Dakshini, K.M.M. and Soni, J.K. (1979), Water quality of sewage drains entering Yamuna in Delhi. Indian J. Environ. Hlth., **21** : 354-360.

Datta, N.C., Bandyopadhyay, B.K., Majumder, A. and Ahuja, D. (1988), Hydrobiological profile of Hooghly sector (Bally to Bandel) of river Ganga. In "Ecology and Pollution of Indian Rivers" (R.K. Trivedy ed.) pp. 113-129. Ashish Publishing House, New Delhi, India.

Dora, M.M. and Roy, N.N. (1987), Investigation of water quality of Subarnarekha river for irrigation. Indian J. Environ. Hlth., ***29*** : 292-298.

George, M.G., Qasim, S.Z. and Siddiqi (1966), A limnological survey of the river Kali with reference to fish mortality. Environ. Hlth., *8* : 262-269.

Geevarghese, C. and Chandrasekharan, M.P. (1985), Studies on the ecological impact of a newsprint factory waste on the river ecosystem. Poll. Res., **4** : 21-24.

Ghose, M.C. and Sharma, C.B. (1988), Effect of drain water on the physico-chemical and bacteriological characters of the river Ganga at Patna, Bihar. In "Ecology and Pollution of Indian Rivers " (R. K. Trivedy ed.) pp. 255-269. Ashish Publishing House, New Delhi, India.

Jagdale, M.H., Salunkhe, M.M., Vibhute, Y.B., Khodaskar, S.N. and Ghuge, K.N. (1984), Pollution of Godavari river water at Nanded. Poll. Res., **3** : 83-84.

Jebanesan, A., Selvanayagam, M. and Thatheyus, A. (1987), Distributory pattern of dissolved oxygen in the selected station of Cooum river and its effect on the aquatic fauna. Proc. Symp. Environ. Biol., 303-311.

Lakshmanarao, N.S. and Narayanarao, M. (1987), Pollution in selected rivers of India–Three case studies. Ind. J. Environ. Stad., **29** : 17-26.

Mahadevan, A. and Krishnaswamy, S. (1983), A quality profile of river Vaigai (S. India). Indian J. Environ. Hlth., **25** : 288-299.

Mishra, G.D., Misra, S., Upadhyaya, S. and Chobey, L.N. (1986), Study of pollution in the Subarnarekha river by the aluminium factory discharge at Muri district Ranchi (Bihar). Ind. Poll. Cont., **2** : 13-16.

Mishra, G.D. and Misra, S. (1987), Chromate pollution in the river Jhoomar district Ranchi, Bihar. J. Ind. Poll. Cont., **3**: 37-40.

Mishra, K.D. (1996), Impact of sewage and industrial pollution on physico-chemical characteristics of water in river Betwa at Vidisha, Madhya Pradesh. In "Assessment of Water Pollution" (S.R. Mishra ed.) pp. 453-463. APH Publishing Corporation, New Delhi, India.

Mishra, S.R. (1990), Hydrobiological characteristics of Morar river, Gwalior in relation to plankton and productivity. Ph.D. Thesis, Jiwaji University, Gwalior (M.P.).

Mitra, A.K. (1982), Chemical characteristics of surface water at a selected gauging station in the river Godavari, Krishna and Tungabhadra. Ind. J. Environ. Hlth., **24** : 165-179.

Mittal, S. and Sengar, R.M.S. (1991), Studies on the distribution of algal flora in polluted regions of Karwan river at Agra (India). In "Current Trends in Limnology-I" (Nalin K. Shastree ed.) pp. 221-230. Narendra Publishing House, Delhi, India.

Nandkumar, N.V., Backyarathy, D.M. and Bhagyalakshmi, K. (1988), Niva river pollution near Chittor town. In "Ecology and Pollution of Indian Rivers" (R.K. Trivedy ed.) pp. 247-254. Ashish Publishing House, New Delhi, India.

Nautiyal, P., Dobriyal, A.K., Chaukiyal, D.C. and Singh, H.R. (1986), Quality and potability of Alkananda water of

Srinagar and Garhwal : A note on bacterial density as an indicator of pollution. Poll. Res., **5** : 143-145.

Nema, R.S. and Lal, B. (1985), Pollution studies of river Sone in Madhya Pradesh. Proc. Nat. Sem. Poll. Cent. Environ. Manag., **3** : 30-44.

Palharya, J.P. and Malviya, S. (1988), Pollution of the Narmada river at Hoshangabad in Madhya Pradesh and suggested measures for control. In "Ecology and Pollution of Indian Rivers" (R.K. Trivedy ed.) pp. 54-85. Ashish Publishing House, New Delhi, India.

Pande, R.K., Rawat, D.S. and Pant, A. (1988), Seasonal rhythm in the physico-chemical properties of Nana Kosi river (Kumaun Himalaya). In "Ecology and Pollution of Indian Rivers" (R.K. Trivedy ed.) pp. 209-218. Ashish Publishing House, New Delhi, India.

Patil, M.R. and Patil, A.R. (1983), Water quality of Ulhas river with respect to copper, cadmium and mercury. Poll. Res., 2: 24-27.

Prakash, C., Rawat, D.C. and Grover, P.F. (1978), Ecological study of the river Jamuna. IAWPC Tech. Annual, *V* : 32-45.

Raina, V., Shah A.R. and Ahmed, S.R. (1984), Pollution studies on river Jhelum I : An assessment of water quality. Ind. J. Environ. Hlth., **26** : 187-201.

Rajagolan, S.B., Dhaneshwar, A.K. and Rao, C.S.G. (1973), Pollution of river Subarnarekha at Ranchi - A survey. Environ. Hlth., **12** : 246-259.

Ramesh, N.B., Razia, S., Reddy, V.K. and Raza, S.H. (1986), A note on characteristics of river Tungabhadra water in Kurnoof and Alampur subjected to paper mill pollution. Acta. Ecol., *8* : 47-50.

Rao, S.P. and Rao, K.S. (1986), Pollution due to pilgrim bathers during Kumbha Mela 1980 in Kshipra river (Ujjain). J. Hydrobiol., **2** : 47-55.

Ray, P. and David, A. (1966), Effect of industrial wastes and sewage upon the chemical and biological composition and fisheries of the river Ganga at Kanpur (U.P.). Environ. Hlth., **8** : 307-339.

Sangu, R.P.S. and Sharma, K.D. (1985), Studies on water

pollution of Yamuna river at Agra. Ind. J. Environ. Hlth., *27* : 257-261.

Saxena, K.L., Chakrabarty, R.N., Khan, A.Q., Chattopadhya, S.N. and Chandra, H. (1966), Pollution studies of the river Ganges near Kanpur. Environ. Hlth., *8* : 270-285.

Shah, A.R. (1988), Physico-chemical aspects of pollution in river Jhelum (Kashmir) during 1981-83. In "Ecology and Pollution of Indian Rivers" (R.K. Trivedy ed.) pp. 163-207. Ashish Publishing House, New Delhi, India.

Shanker, V., Sangu, R.P.S. and Joshi, G.C. (1986), Impact of distillery effluent on the water quality and ecosystem of river Reh in Doon valley. Poll. Res., *5* : 137-142.

Sharma, K.D., Lal, N. and Pathak, P.D. (1981), Water quality of sewage drains entering Yamuna at Agra. Indian J. Environ. Hlth., **23** : 118-122.

Sharma, R.C. (1984), Potamological studies on lotic environment of the upland river Bhagirathi of Garhwal Himalaya. Environ. & Ecol., **2** : 239-242.

Sikander, M.H.D. and Tripathi, B.D. (1983), The physico-chemical characteristics of Ganga water at Varanasi. Proc. Natl. Conf. River Poll. Human Hlth., Varanasi, 53-62.

Singh, A.K. (1997), Abundance of macrozoobenthic organisms in relation to physico-chemical characteristics of river Ganga at Patna (Bihar), India. J. Environ. Biol., **18**(2) : 103-110.

Singh, D.K. and Singh, C.P. (1990), Pollution studies on river Subernarekha around industrial belt at Ranchi (Bihar). Indian J. Environ. Hlth., **32** : 26-33.

Singh, H.P. and Mahaver, L.R. (1997), Preliminary observations on heavy metals in water and sediments in a stretch of river Ganga and some of its tributaries. J. Environ. Biol., ***18(1)*** : 49-53.

Somashekar, R.K. (1984), Studies of water pollution of the river Cauvery–Physico-chemical characteristics. Intern. J. Environ. Studies, **23** : 209-216.

Sreenivasan, A. and Ray, R.S. (1967), Effect of certain wastes on the water quality and fisheries of river Cauvery and Bhavani Environ. Hlth., ***9*** : 13-21.

Sunder, S. (1988), Monitoring of water quality in a stretch of river Jhelum, Kashmir. In "Ecology and Pollution of Indian Rivers" (R.K. Trivedy ed.) pp. 131-161. Ashish Publishing House, New Delhi, India.

Unni, K.S., Chauhan, A., Varghese, M. and Naik, L.P. (1992), Preliminary hydrobiological studies of river Narmada from Amarkantak to Jabalpur. In "Aquatic Ecology" (Mishra, S.R. and Saksena, D.N. eds.) pp. 221-229. Ashish Publishing House, New Delhi, India.

Vass, K.K., Raina, H.S., Zutshi, D.P. and Khan, M.A. (1977), Hydrobiological studies on river Jhelum. Geobios, **4** : 238-242.

Venkateswarlu, V. (1969), An ecological study of the algae of the river Moosi, Hyderabad (India) with special reference to water pollution. I physico-chemical complex. Hydrobiologia, **33** : 117-143.

Verma, S.R. and Dalela, R.C. (1975), Studies on the pollution of the Kalinadi by industrial wastes near Mansurpur, Part-I : Hydrometric and physico-chemical characteristics of the wastes and river water. Acta. Hydrochim. Hydrobiol., 3: 239-257.

Verma, S.R., Tyagi, A.K. and Dalela, R.C. (1978), Physico-chemical and biological characteristics of Kadarabad in Uttar Pradesh. Indian J. Environ. Hlth. ***20(1)*** : 1-13.

Verma, S.R., Shukla, G.R. and Dalela, R.C. (1980), Studies on the pollution of Hindon river in relation to fish and fisheries. Limnologica (Berlin), **12** : 33-75.

Verma, S.R., Sharma, P., Tyagi, A., Rani, S., Gupta, A.K. and Dalela, R.C. (1984), Pollution and saprobic status of Eastern Kalinadi. Limnologica (Berlin), **15** : 69-133.

12

Studies of the Benthic and Shore-Line Fauna of River Ganga Between Narora to Kannauj : A Comprehensive Study

Asif A. Khan, Rajeev K. Gaur & Md. Afaque Alam

INTRODUCTION

The community of organisms, living on the bottom of a waterbody, called benthos and form an important link in the food web of fishes apart from their biological role in the mud-water exchange of nutrients. In recent years, the growing recognition for better understanding of benthic environment, it's communities and productivity, has lead to increased exploitation of many inland waterbodies (Rosillon, 1985; Cohen, 1986 and Bass, 1986).

Review of literature on the subject reveal that most of the studies are from fish ponds whereas such studies for rivers specially on Ganga are very scanty except a few important contributions made by Pahwa (1979) and Jhingran (1988) in different stretches of the river. In view of the scanty information over the subject in river Ganga, the present investigations were conducted.

Attempts were made to examine the comparative account of distribution and abundance of benthic and shore-line fauna of river Ganga in 236 km virgin stretch between Narora to Kannauj, in Uttar Pradesh.

MATERIALS AND METHODS

The benthic and shore-line fauna were collected seasonally from 1987 to 1990 from four stations of river Ganga. The criteria of selection of the sites were approachability by vehicle and as far as possible a uniform inter-station distance.

Samples were collected by Ekman dredge and transferred in a 15 litre plastic buckets containing some water. After complete mixing of water and mud samples by stirring, the suspension was seived-out through metallic seives having 2 mm and 0.5 mm meshes. The macro-invertebrates were picked-up by forceps and transferred in a bottle containing 70% alcohol and the smaller microscopic benthic fauna were obtained by floatation, adding 5 gm of sucrose to the residual organic matter retained in the seive with 50 to 100 ml of water. Shore line fauna was collected manually from the banks of the river. These samples were then identified following the works of Edmondson (1959), Needham and Needham (1962) and Pennak (1978).

RESULTS AND DISCUSSION

Seasonal distribution and abundance of benthic and shore-line fauna of various sampling stations are presented in Tables 1-4 and summarized in Table 5. The investigations indicate well defined fluctuations in the distribution and abundance of various groups of benthos during different seasons at different sampling stations. Whereas, Cowell and Vodopich (1981) reported an uniformity in the abundance of oligochaetes throughout the year.

The major groups of benthos and shore-line fauna belong to six major taxa, namely Bivalvia, Gastropoda, Ostracoda, Oligochaeta, Insecta and Nematodes. Out of these, the most abundant group at all the sampling stations was found to be insecta followed by gastropoda, bivalvia, ostracoda and oligochaeta except at station Katchla, where instead of insecta, gastropoda was most abundant. Nematodes were also observed in fairly high numerical strength although no attempt was made to identify various species of nematodes.

On comparing benthic and shore-line fauna at different stations, it is observed that their abundance was in decreasing order as Narora > Katchla > Kannauj > Fatehgarh. The maximum abundance of benthos at Narora might be due to receiving large quantity of nutrients and organic matter as a result of throwing half burnt human dead-bodies at Rajghat, a station few km above Narora in the upstream alongwith worship and holy bathings at Narora which might have

TABLE 1 : SEASONAL VARIATIONS IN BENTHIC AND SHORE-LINE FAUNA AT NARORA (1987-90)

Genera	PMON '87	WINT '87	SPRG '88	SUMM '88	PMON '88	WINT '88	SPRG '89	SUMM '89	PMON '89	WINT '89	SPRG '90
BIVALVIA											
Carbicula sp.	–	+	+	++	+++	+++	±	+	++	+++	–
Indonia sp.	–	+	–	+++	++	++	–	+	–	++	–
Lamelidens sp.	–	+	+	+	–	–	+	+	–	–	–
Unio sp.	–	–	–	–	–	–	–	–	–	–	–
Pisidium sp.	–	–	–	–	+	+	–	–	+	+	–
GASTROPODA											
Bithynia sp.	–	–	–	–	–	–	–	++	–	+	–
Gyrulus sp.	–	–	–	–	–	–	+	+++	–	+	++
Indoplanorbis sp.	–	–	+	+	–	–	+	+	–	+	+
Lymnea sp.	–	–	–	+	–	–	++	+++	–	–	++
Melania sp.	+	+	+	–	–	–	+	–	+	–	+
Vivipara bengalensis	–	+	+	+	++	++	+	++	–	+	++
Subulina sp.	–	++	++	+++	++	++	+++	+++	–	++	++
OSTRACODA											
Cypris sp.	–	–	+	+	–	–	±	+	–	–	±
Centrocypris sp.	–	–	+	–	–	–	+	–	–	–	++
Heterocypris sp.	+++	++	+++	+	+	–	+++	±	++	+	+++
Stenocypris sp.	–	–	–	–	–	–	–	–	–	–	–
Ostracod shell	–	–	–	–	–	–	–	–	–	–	–

OLIGOCHAETA											
Tubifex	++	++	++	+	±	–	+++	+	+	–	++
INSECTA											
Chironomus sp.	+	++	+++	+	++	++	+++	++	+++	+++	+++
Culicoides sp.	+	–	+	–	++	++	+	–	+	++	++
Caenis (nymph)	–	–	±	–	+	±	+	–	–	+	±
Megistocera sp.	–	–	–	–	–	–	–	–	–	–	–
Simulium sp.	–	–	–	–	+	+	–	–	–	–	–
Stenelemis sp.	+		–	–	–	–	–	–	–	–	–
Tendipes sp.	++	+	+	+	+++	+++	±	+	++	+	±
Chaoborus sp.	–	–	–	–	++	–	–	–	–	–	–
Polypedilium sp.	–	–	–	–	+	–	–	–	–	–	–
NEMATODES	+++	++	++	++	+	+	++	++	+	++	+++

ABUNDANCE :

– Absent ± Rare + Present

++ Abundant +++ Highly abundant

TABLE 2 : SEASONAL VARIATIONS IN BENTHIC AND SHORE-LINE FAUNA AT KATCHLA (1987-90)

Genera	PMON '87	WINT '87	SPRG '88	SUMM '88	PMON '88	WINT '88	SPRG '89	SUMM '89	PMON '89	WINT '89	SPRG '90
BIVALVIA											
Carbicula sp.	–	–	+	+	++	+	+	+	+++	++	+++
Indonia sp.	–	+	+	+	++	++	±	+	–	+++	+
Lamelidens sp.	–	–	–	–	–	–	–	–	±	±	–
Unio sp.	–	–	+	+	–	–	–	–	–	–	–
Pisidium sp.	–		–		±	+	–	–	±	+	–
GASTROPODA											
Bithynia sp.	–	–	+	+	++	++	–	+	–	+	–
Gyrulus sp.	–	–	–	–	++	++	–	±	–	++	–
Indoplanorbis sp.	–	–	–	–	++	++	–	±	+++	+++	–
Lymnea sp.	–	–	–	–	–	–	–	++	–	+	+
Melania sp.	–	–	–	–	–	–	–	–	±	–	–
Subulina sp.	–	++	++	+++	++	++	+++	+++	–	++	++
Vivipara bengalensis	–	–	–	–	+++	+++	++	–	±	+++	+++
OSTRACODA											
Cypris sp.	+	–	++	–	–	–	–	–	+	+	++
Centrocypris sp.	–	–	–	–	–	–	–	–	–	–	–
Heterocypris sp.	+++	++	+	+	+	–	+	+	+	+	+
Stenocypris sp.	–	–	–	–	–	–	–	–	–	–	–
Ostracod shell	–	–	–	–	–	–	–	–	–	++	–

OLIGOCHAETA											
Tubifex	–	–	–	–	++	++	–	–	++	+++	+
INSECTA											
Chironomus sp.	–	+	±	+	++	++	+	+	++	++	+
Culicoides sp.	±	+	–	–	++	++	±	–	++	+++	±
Chaoborus sp.	–	–	–	–	+	–	–	–	–	–	–
Caenis (nymph)	+		–	–	–	–	–	–	+	–	–
Megistocera sp.	–	+	–	–	+	±	–	–	–	+	–
Polypedilium sp.	–	–	–	–	++	–	–	–	–	–	–
Simulium sp.	–	–	–	–	–	–	–	–	–	+	–
Steneiemis sp.	–	–	–	–	–	–	–	–	–	–	–
Tendipes sp.	+	+	–	–	–	–	–	±	–	+	–
NEMATODES	+++	+	++	+	+++	+++	++	+	+++	+++	++

ABUNDANCE :

– Absent **±** Rare **+** Present

++ Abundant **+++** Highly abundant

TABLE 3 : SEASONAL VARIATIONS IN BENTHIC AND SHORE-LINE FAUNA AT FATEHGARH (1987-90)

Genera	PMON '87	WINT '87	SPRG '88	SUMM '88	PMON '88	WINT '88	SPRG '89	SUMM '89	PMON '89	WINT '89	SPRG '90
BIVALVIA											
Carbicula sp.	+	+	–	–	+++	++	–	–	++	+	
Indonia sp.	–	–	–	–	+	++	–	±	–	++	
Lamelidens sp.	+++	++	++	++	++	++	++	–	++	++	
Unio sp.	–	–	–	–	–	–	–	+	–	–	
Pisidium sp.	–	–	–	–	–	–	–	–	–	–	
GASTROPODA											
Bithynia sp.	–	–	–	+	–	–	–	+	–	–	
Gyrulus sp.	–	–	–	–	–	–	–	–	–	–	
Indoplanorbis sp.	+	+	–	–	±	+	–	–	+	±	
Lymnea sp.	–	–	–	–	–	–	+	+	–	–	
Melania sp.	–	–	–	–	–	–	–	±	+	–	
Subulina sp.	–	+	+	+	–	+++	+	+	–	+	
Vivipara bengalensis	+	+	++	+	–	–	++	+	+	–	
OSTRACODA											
Cypris sp.	–	–	–	–	–	–	–	–	–	–	
Centrocypris sp.	–	–	–	–	–	–	–	–	–	–	
Heterocypris sp.	++	+	–	+++	–	–	±	+++	±	–	
Stenocypris sp.	–	–	±	–	–	–	–	–	–	–	
Ostracod shell											

OLIGOCHAETA										
Tubifex	–	+	–	–	+++	++	–	–	++	+
INSECTA										
Chironomus sp.	+	++	++	+	++	++	++	+	+++	++
Culicoides sp.	+	+	+	–	++	++	+	–	±	++
Caenis (nymph)	–	–	–	–	+	±	–	–	–	–
Megistocera sp.	–	–	–	–	++	++	–	–	–	+
Simulium sp.	–	–	–	–	–	–	–	–	–	–
Stenelemis sp.	–	–	–	–	+	–	–	–	–	–
Tendipes sp.	+	+	+	+	++	++	+	–	++	–
NEMATODES	±	+	±	++	+++	+++	+	++	++	++

ABUNDANCE :

– Absent ± Rare + Present

++ Abundant +++ Highly abundant

TABLE 4 : SEASONAL VARIATIONS IN BENTHIC AND SHORE-LINE FAUNA AT KANNAUJ (1987-90)

Genera	PMON '87	WINT '87	SPRG '88	SUMM '88	PMON '88	WINT '88	SPRG '89	SUMM '89	PMON '89	WINT '89	SPRG '90
BIVALVIA											
Carbicula sp.	+++	++	+++	–	+++	+++	+++	–	+++	+++	+++
Indonia sp.	–	–	+++	+++	–	–	++	++	–	–	–
Lamelidens sp.	–	+	++	+	–	–	+	±	–	–	–
Unio sp.	–	–	–	++	–	–	–	++	–	–	–
Pisidium sp.	–	–	–	–	–	–	–	–	–	–	–
GASTROPODA											
Bithynia sp.	–	–	++	+	+++	+++	–	+	–	++	–
Gyrulus sp.	–	–	+	–	+	+	–	–	–	+	–
Indoplanorbis sp.	++	++	+++	++	+++	+++	++	++	+++	+++	+
Lymnea sp.	–	–	+++	++	–	–	–	++	–	–	–
Melania sp.	–	–	–	–	–	–	–	–	++	–	–
Subulina sp.	–	++	–	–	+	+	–	–	–	++	+
Vivipara bengalensis	++	++	+	+	++	++	+	+	++	++	+
OSTRACODA											
Cypris sp.	+	–	–	–	–	–	–	–	++	+	–
Centrocypris sp.	–	–	–	–	–	–	–	–	–	+	±
Heterocypris sp.	+++	++	–	–	–	–	–	–	–	–	±
Stenocypris sp.	–	–	–	–	–	–	–	–	–	–	–
Ostracod shell	–	–	–	+	+	–	–	+	–	–	–

OLIGOCHAETA											
Tubifex	–		–	–	+++	++	–	–	++	+++	±
INSECTA											
Chironomus sp.	++	+	+	+	+++	+++	+	+	++	+++	+
Culicoides sp.	±	++	+	–	+++	+++	+	–	+++	++	–
Caenis (nymph)	±	–	+	+	++	++	–	–	±	++	+
Megistocera sp.	–	++	–	–	+	+	–	–	–	±	–
Simulium sp.	–	–	–	–	–	–	–	–	–	–	–
Stenelemis sp.	–	–	–	–	–	–	–	–	–	–	–
Tendipes sp.	±	–	–	–	–	–	–	–	±	+	–
Tipula	–	–	–	–	–	++	–	–	–	–	–
NEMATODES	+++	++	+++	+	++	++	+++	+	+	++	++

ABUNDANCE :

– **Absent** ± **Rare** + **Present**

++ **Abundant** +++ **Highly abundant**

TABLE 5 : ABUNDANT BENTHOS AND SHORE-LINE FAUNA (1987-90)

Season	Stations			
	Narora	Kachla	Fategarh	Kannauj
PMON '89	*Corbicula* sp. *Hetrocypris* sp. *Chironomus* sp. *Tendipes* sp.	*Indoplanorbis* sp. *Corbicula* sp. *Chironomus* sp. *Culicoides* sp. *Tubifex* sp. Nematodes	*Corbicula* sp. *Chironomus* sp. *Tendipes* sp. *Tubifex* sp. Nematodes	*Indoplanorbis* sp. *Melania* sp. *V. bengalensis* *Corbicula* sp. *Chironomus* sp. *Culicoides* sp. *Tubifex*
WINT '89–90	*Subulina* sp. *Corbicula* sp. *Indonia* sp. *Chironomus* sp. *Culicoides* sp. Nematodes	*Gyrulus* sp. *Indoplanorbis* sp. *Subulina* sp. *V. bengalensis* *Corbicula* sp. *Indonia* sp. *Chironomus* sp. *Culicoides* sp. *Tubifex* Nematodes	*Indonia* sp. *Lamellidens* sp. *Chironomus* sp. *Culicoides* sp. Nematodes	*Bithynia* sp. *Indoplanorbis* sp. *Subulina* sp. *V. bengalensis* *Corbicula* sp. *Chironomus* sp. *Culicoides* sp. *Tubifex*
PWINT '90	*Gyrulus* sp. *Lymnea* sp. *Subulina* sp. *V. bengalensis* *Corbicula* sp. *Indonia* sp. *Centrocypris* sp. *Hetrocypris* sp. *Chironomus* sp. *Culicoides* sp. *Tubifex* Nematodes	*Subulina* sp. *V. bengalensis* *Corbicula* sp. *Cypris* sp. Nematodes	ND	*Corbicula* sp. Nematodes

encouraged the multiplication of benthos. Whereas, on comparing total benthic and shore-line fauna during different seasons, it was observed that their abundance was maximum either in post-monsoon or in winter. This may be attributed to the fact that the rains, surface run-off and drainage during monsoon has added lot of organic matter and nutrients to the river, thereby, encouraging the multiplication of benthos (Olive and Dambach, 1973). Whereas, Bais *et al*. (1992) correlated maximum abundance of benthos with the release of untreated sewage into the aquatic environment. The maximum abundance and distribution of benthic and shore-line fauna during winter season in the present investigation might be due to more availability of organic matter during winter seasons. Mandal and Moitra (1975) also reported peaks of benthos in winter and early summer months.

A total of 28 species of benthic and shore-line fauna belonging to groups Bivalvia, Gastropoda, Ostracoda, Insecta & Oligochaeta, were observed. No attempt was made to identify nematodes. From Table 1-4 it is also evident that a few species were found absent during certain seasons and reappeared during others. This might be due to return of favourable conditions which enhanced the growth and reproduction of those species. Haque (1991) and Gaur (1994) also reported similar findings in their studies on plankton organisms.

On comparing different stations, it was observed that certain species were absent at one station while present at other station. This observation indicates the nature of the river bed at that station as benthos are indicators of water quality (Olive and Dambach, 1973). Yet the population of the benthic and shore-line fauna was found to be stable, perhaps because of not much variation in the ecological conditions of different stations. Barbhuyan and Khan (1992) also observed similar results in their studies on benthos in ponds.

MAJOR TAXA

1. *Mollusca* : These were represented by bivalvia and gastropoda.

   (a) *Bivalvia* : A total of 5 members of bivalvia were observed. Out of these, *Corbicula* sp. and *Indonia* sp. were abundant at all the four sampling stations. Besides these, *Lamelidens* sp. was noted to be in maximum abundance during all the seasons at Fatehgarh.

(b) *Gastropoda* : Although a number of species were encountered belonging to gastropoda but the most noted during all the seasons at all the sampling stations were *Vivipara bengalensis* and *Subulina* sp. At Kannauj, besides these, *Indoplanorbis* sp. was also found to be most abundant during all the seasons.

2. *Ostracoda* : This group was represented by 5 species including the most dominant, *Hetrocypris* sp, occurring in all the samples and at all the sampling stations. However, *Cypris* sp. was markedly absent at Fatehgarh.
3. *Oligochaeta* : This group was represented by single member, *Tubifex* sp., occurring abundantly at Narora and least at Fatehgarh. As oligochaetes, particularly *Tubifex,* are inhabitants of mud enriched with organic matter (Wilham and Dorris, 1966 and Adholia *et al.* 1990), station Narora seems to be organically polluted as compared to other stations. The probable sources of this organic matter at Narora are the throwing of half burnt human dead-bodies at Rajghat, a station few km in the upstream of Narora, and addition of night soil in large amounts by the people worshipping and bathing at Narora regularly during all the seasons.
4. *Insecta* : This was the most abundant group observed during the present investigations, represented by 9 genera. Among them, the most important species sampled during all the seasons were *Chironomus* sp. and *Culicodes* sp., whereas, *Megistocera* sp., *Simulium* sp. and *Stenelium* sp. were less abundant and were rarely encountered during the present investigations. Spatially, the most abundant species were *Chironomus* sp. and *Tendipes* sp. at Narora followed by Fatehgarh, Kannauj and Katchla. Presence of *Chironomus* sp. and *Tendipes* sp., which are indicators of organically enriched habitats (Wilham and Dorris, 1966 and Adholia *et al,* 1990), also reveals that Narora is the most polluted point organically.
5. *Nematodes* : These were also observed in fairly high numerical strength at all the sampling stations during all the seasons.

CONCLUSION

From the above discussion, it can be concluded that insects and molluscs regulate the seasonal and stationwise variations of the benthic and shore line fauna. The presence of indicator species, like *Tubifex, Chironomus* and *Tendipes* in most abundance, at Narora, indicates the high trophic status of the station. The organic matter also seemed to be the main limiting factor which controlled the distribution pattern of benthic and shore-line fauna.

REFERENCES

Adholia, U.N., Chakrawarti, A., Srivastava, V. and Vyas, A. (1990). Community studies on macro-zoobenthos with reference to limno-chemistry of Mansarovar Reservoir, Bhopal *J. Natcon.* **2** (2) : 139-154.

Bais, V.S., Yatheesh, S. and Agrawal, N.C. (1992). Benthic macro-invertebrates in relation to water and sediment chemistry. *J. Freshwater Biol.* **4**(3) : 183-191.

Barbhuyan S.I. and Khan, A.A. (1992). Studies on the structure and function of benthic ecosystem in a eutrophic body of water : Temporal and spatial distribution of benthos. *J. Freshwater Biol.* **4**(4) : 239-247.

Bass, D. (1986). Larval Chironomids (Diptera) of the Big Thicket Streams. *Hydrobiologia,* **135**: 271-285.

Cohen, A.S, (1986). Distribution and faunal association of benthic invertebrates of Lake Turkana, Kenya, *Hydrobiologia,* **141** (3): 179-197.

Cowell, B.C. and Vadopich, D.S. (1981). Distribution and seasonal abundance of benthic macro-invertebrates. *Hydrobiologia, 78* (2) : 97-105.

Edmondson, W.T. (1959). *Freshwater Biology* (2nd Edn.) John Wiley & Sons. Inc. New York, p. 1248.

Gaur, R.K. (1994). Temporal and Spatial distribution of cyanobacteria in a polluted water body. M.Phil. Dissertation, Aligarh Muslim University, Aligarh. 62 p.

Haque, N. (1991). Studies on the hydrobiology of some polluted ponds of Aligarh region. Ph.D. Thesis, Aligarh Muslim University, Aligarh. 219 p.

Jhingran, V.G. (1988). *Fish and Fisheries of India* (2nd Edn.) Hindustan Publishing Corporation, New Delhi. 666 p.

Mandal, B.K. and Moitra, S.K. (1975). Studies on the bottom fauna of a freshwater fish pond at Burdwan *J. Inland Fish. Soc. India*, *7* : 43-48.

Needham, J.G. and Needham, P.R. (1962). *A guide to the freshwater Biology* (5th Edn.). Holden day Inc., San Francisco : 108 p.

Olive, H.J. and Dambach, C.A. (1973). Benthic macro-invertebrates as indicators or water quailty in wheatstone creek, Marrow country, (Ohio Scioto River Basin). *Ohio J. Sci.* **73** (3) : 129-149.

Pahwa, D.V. (1979). Studies on the distribution of the benthic macrofauna in the stretch of River Ganga. *Indian J. Anim. Sci.* **49** (3) : 212-219.

Pennak, R.W. (1978). *Freshwater Invertebrates of the United States.* John Wiley & Sons, New York. 803 p.

Rosillon, D. (1985). Seasonal variations in the benthos of a chalk trout stream, the river Samson, Belgium. *Hydrobiologia* **126** : 253-262.

Wilhm, R.L. and Dorris T.C. (1966). Species diversity in benthic macro-invertebrates in stream receiving domestic and oil refinery effluent. *Am. Midl. Nat.* **76** : 427-449.

13

Seasonal Trends in Abiotic Factors of a Lentic Habitat (Kalika Pond, Dhar)

R.K. Dave, M.M. Prakash & N.K. Dhakad

INTRODUCTION

Abiotic factors make up the basic environmental structure. Each of these factors as their units act singly or in syncronized manner, exirts significant influence upon the aquatic life. Hence these are important tools to assess the water quality and improve the productivity of lentic habitats.

Now-a-days increased human population and industrial development has put tremendous pressure on the limited fresh water resources (lotic and lentic). Thus it is our primary need to investigate fresh water habitats as many as we can for their proper use and management.

The district Dhar (M.P.) has many lotic and lentic habitat. However, no attempt has been made to investigate them, though they have good potential for aquaculture and pisciculture. Present paper thus attempts to analyse abiotic factors of a lentic habitat (Kalika Pond, Dhar) for proper management and use.

MATERIALS AND METHODS

The lentic habitat selected for the present investigation is known as Kalika Pond. The Kalika Pond is situated in the township of district Dhar (M.P.). It is geographically located between latitude, 22°35' North and longitude 75°20' East, at a height of 592 meters above the mean sea level. The pond is supported by black cotton soil (vegur). Its morphometry was described by Dave and Shinde (1986).

Water samples for the study of abiotic factors was collected monthly from April 1986, to March 1991. Analysis of various abiotic factors was done by according to method given by APHA (1985) and Trivedi (1986).

RESULTS AND DISCUSSION

The monthly data obtained for the various abiotic factors are presented in Table-1a, b & c. The distribution pattern are shown in the Fig. 1 and 2.

Results obtained indicate that the surface water temperature in Kalika Pond ranged from 16.2° to 30.4°C. in 1989-90 and from 14.0°C to 30.2° in 1990-91. The variation in the annual range was 14.2°C in 1989-90 and 16.2°C in 1990-91 (Table-1a). This range of variation was found higher than those reported by Ganapati (1960), Zafar (1966), Jana & Sarkar (1971), Bohra & Bhargava (1977) and Prakash (1994). In the present investigation during summer, rise in temperature was noted which was due to bright sun light and humidity in the air. In the monsoon months lowering in temperature was noted. It was due to the reduced solar radiation and the presence of cloudy weather. During the winter season the temperature remained low due to cold, low ambient temperature and shorter photoperiod.

Verma & Shukla (1970), Hannan (1979), Qadri *et al.* (1981), Gupta & Mehrotra (1986), Yadav *et al.* (1987), Shastree *et al.* (1991) and Prakash (1996) observed that at high temperature the dissolved oxygen was low, while Hussainy (1965) noted that at low temperature, the amount of oxygen is always low. However, Sreenivasan (1974) reported that water temperature did not affect the dissolved oxygen. In the present study in June when the water temperature was maximum, the value of dissolved oxygen was lowest, while in December when the water temperature was low the values of dissolved oxygen were fairly high. Thus it had an inverse relationship with the dissolved oxygen.

The pH is a measure of the acid-base equilibrium achieved by various dissolved compounds and is controlled by the carbon-di-oxide, carbonate and bi-carbonate equilibrium system in most natural water (Kaushik *et al.* 1991). In the present investigation water of the Kalika Pond always exhibit an alkaline range of pH varying between 7.4 to 9.3 in 1989-90 and 7.4 to 9.0 in 1990-91 (Table-1b). Sreenivasan (1979), Rao *et al.* (1977), Shrivastava *et al.* (1977), Nayak *et al.* (1982), Nair *et al.*

TABLE 1(a) : SEASONAL VARIATIONS IN THE PHYSICAL CONDITIONS OF KALIKA POND WATER AT SURFACE

Sl. No.	Month	Season	Water Level (m)		Colour		Temperature (°C)		Secchi disc transparency (cm)	
			1989-90	1990-91	1989-90	1990-91	1989-90	1990-91	1989-90	1990-91
1.	April	Summer	1.00	1.28	Light green	Light green	25.5	26.2	30	29
2.	May		0.76	1.00	Light green	Light green	28.4	28.0	28	27
3.	June		0.61	0.88	Dusky	Dusky	30.4	30.2	20	19
4.	July	Monscon	1.00	1.50	Dusky	Dusky	28.5	24.2	10	8
5.	August		4.00	4.00	Dusky	Dusky	26.7	23.5	8	7
6.	September		4.00	4.00	Dusky	Dusky	27.7	27.5	17	19
7.	October		3.48	3.90	Green	Green	27.8	28.0	84	80
8.	November		3.00	3.50	Dark green	Dark green	22.5	22.0	87	89
9.	December	Winter	3.00	3.00	Dark green	Dark green	16.2	14.0	74	76
10.	January		2.48	2.68	Dark green	Dark green	21.1	20.2	56	58
11.	February		2.38	2.58	Light green	Light green	22.5	21.5	43	45
12.	March	Summer	2.00	2.28	Light green	Light green	23.7	23.0	33	37

TABLE 1 (b) : SEASONAL VARIATIONS IN THE CHEMICAL CONDITIONS OF POND WATER AT SURFACE

Sl. No.	Month	Season	Dissolved Oxygen (ppm)		Hydrogen-ion concen. (pH)		Free carbon dioxide (ppm)		Carbonates (ppm)		Bicarbonates (ppm)		Total alkalinity (ppm)	
			1989-90	1990-91	1989-90	1990-91	1989-90	1990-91	1989-90	1990-91	1989-90	1990-91	1989-90	1990-91
1.	April	Summer	7.3	7.1	9.3	9.0	Nil	Nil	22	24	140	135	162	159
2.	May		7.0	7.0	9.0	8.9	Nil	Nil	18	19	160	162	178	181
3.	June		6.4	6.5	8.5	8.2	Nil	Nil	17	18	170	175	187	193
4.	July	Monsoon	6.5	6.7	7.6	7.5	Nil	Nil	15	15	180	185	195	200
5.	August		6.7	6.8	7.4	7.4	Nil	Nil	10	8	190	197	200	205
6.	September		7.2	7.3	8.1	8.2	Nil	Nil	12	13	180	186	192	199
7.	October		7.5	7.7	8.3	8.3	Nil	Nil	13	15	172	175	185	200
8.	November		8.6	8.7	8.4	8.5	Nil	Nil	15	17	165	172	180	189
9.	December	Winter	9.3	9.6	8.2	8.4	Nil	Nil	16	18	160	165	176	183
10.	January		7.8	8.0	8.4	8.3	Nil	Nil	18	20	156	158	174	178
11.	February		7.6	7.7	8.5	8.6	Nil	Nil	20	22	150	150	170	172
12.	March	Summer	7.5	7.8	8.7	8.7	Nil	Nil	21	23	145	140	166	163

TABLE 1 (c) : SEASONAL VARIATIONS IN THE CHEMICAL CONDITIONS OF POND WATER AT SURFACE

Sl. No.	Month	Season	Calcium (ppm)		Chloride (ppm)		Nitrate (ppm)		Inorganic phosphate (ppm)		Silicate (ppm)	
			1989-90	1990-91	1989-90	1990-91	1989-90	1990-91	1989-90	1990-91	1989-90	1990-91
1.	April	Summer	26	24	38	36	0.15	0.15	0.017	0.015	3.6	3.5
2.	May		25	23	39	37	0.15	0.14	0.014	0.013	3.0	3.0
3.	June		28	27	45	43	0.22	0.23	0.025	0.022	4.8	4.1
4.	July	Monsoon	29	30	28	27	0.27	0.29	0.048	0.049	6.1	6.0
5.	August		39	40	25	24	0.39	0.40	0.052	0.054	8.2	8.3
6.	September		37	37	24	23	0.35	0.37	0.046	0.048	7.9	8.0
7.	October		35	34	26	24	0.30	0.33	0.044	0.045	7.2	7.0
8.	November		33	31	29	27	0.28	0.29	0.033	0.035	6.0	5.9
9.	December	Winter	31	30	30	29	0.21	0.20	0.025	0.027	5.4	5.5
10.	January		30	28	32	30	0.19	0.19	0.023	0.025	4.5	4.5
11.	February		29	27	35	32	0.18	0.17	0.020	0.021	4.2	4.2
12.	March	Summer	27	26	35	33	0.16	0.16	0.018	0.020	3.8	4.0

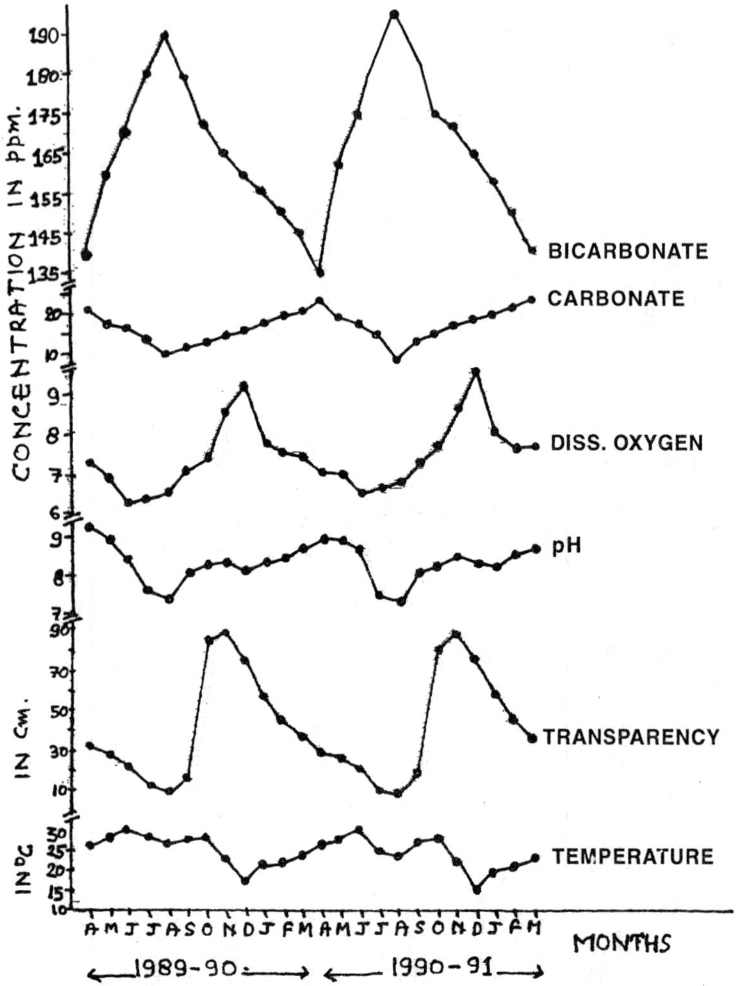

FIG. 1: MONTHLY VARIATION IN PHYSICAL AND CHEMICAL FACTORS

(1988), Prakash (1996) and Rajput and Bhadoria (1996) has also reported alkaline nature of lentic habitat studied by them. In the present investigation the pH was lowest in August (Rainy month) and highest in April (Summer month). These observations corroborate with the observation of Vyas (1968) and Prakash (1994).

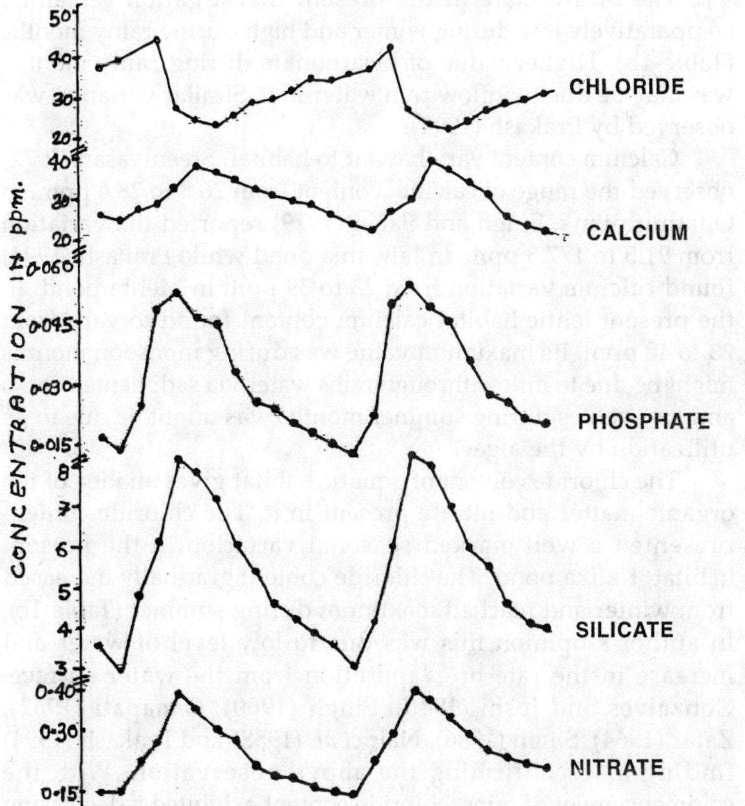

FIG. 2 : MONTHLY VARIATION IN CHEMICAL FACTORS

The carbonate alkalinity of Kalika Pond ranged between 10 and 22 ppm in 1989-90 and between 8 and 24 ppm in 1990-91. It was minimum during the rainy season and maximum during winter and summer. The carbonate alkalinity showed a direct positive correlation with pH as they mostly tended to rise and fall together. This observation is similar to that of Zafar (1964), Sinha (1969) and Singh and Sahai (1979) and Prakash (1996).

The bicarbonate in the present investigation remained comparatively low during winter and high during rainy months (Table-1b). Higher value of bicarbonate during rainy months was may be due to inflow rain water in it. Similar variation was observed by Prakash (1994).

Calcium content vary habitat to habitat. Sreenivasan (1972) observed the range of calcium content from 26.8 to 28.4 ppm. in Odathuraitank. Singh and Sahai (1979) reported the variation from 91.5 to 177.3 ppm. in Jalwania pond while Prakash (1994) found calcium variation from 23 to 39 ppm in Mehta pond. In the present lentic habitat calcium content found to vary from 23 to 40 ppm. Its maximum value was during monsoon months might be due to inflow through rains water via sedimented rocks and low values during summer months was might be due to its utilization by the algae.

The chloride content of aquatic habitat gives an idea of the organic matter and nitrate present in it. The chloride content presented a well marked seasonal variation in the present habitat, Kalika pond. The chloride content gradually increased from winter and reached maximum during summer (Table-1c). In author's opinion this was due to low level of water and increase in the rate of evaporation from the water surface. Gonzalves and Joshi (1946), Singh (1960), Ganapati (1962), Zafar (1964), Singh (1986), Nair *et al.* (1988) and Prakash (1994) finding also confirming the above observation. With the commencement of rains, chloride content exhibited a decreasing trend, which agrees with the result of Ganapati (1962) and Prakash (1994) who concluded the reductions of chlorides during rains due to dilution of water and lower evaporation rate.

The most important inorganic nitrogen compounds in water are nitrate. In the present study the quantity of nitrate decreased from postmonsoon to summer (Table-1c). This may be due to the rapid uptake of nitrates by the phytoplankton. In our opinion this may be due to the rapid uptake of nitrates by the phytoplankton (Tukahashi and Saijo, 1981 and Toetz, 1981) and higher aquatic plants (Schwoerbel and Fillmans, 1977, Metzer and Exier, 1982). Zafar (1964), Venkateshwarlu (1969) and Prakash (1996) noted a positive correlation between the nitrates and dissolved oxygen. In the present study, such relationship was noted in summer and monsoon months. But in the post-monsoon and winter months this relationship was negative. It was due to the higher primary productivity rate yielding a large quantity of dissolved oxygen. However, the nitrates were affected adversely as these were consumed by the phytoplankton. Singh (1960) assumed that fluctuation in nitrates are independent of oxygen. However, Munawar (1970) showed an inverse relationship between nitrates and dissolved oxygen.

Phosphorous recognized as critical single factor for maintaining fertility of lentic habitat. In the present investigation, the dissolved inorganic phosphate showed a well marked seasonal variation. Its maximum values were recorded in the rainy season, while minimum values were noted in the summer season (Table-1c). In the monsoon months, the rise of phosphate level may be attributed to the rains which had also been reported by Hutchinson (1957), Michael (1969), Munawar (1970), Singh and Sahai (1979) and Nair *et al.* (1980). In the summer months, the lower values of inorganic phosphate were due to its utilization by the phytoplankton, which support Kimmel and Lind (1977) view that phytoplankton production is limited by low phosphate availability.

In the present study, fairly large quantities of silicates were observed. The quantity of silicates increased from late summer to post-monsoon, reaching considerably high in August-September, which might be due to inflow of rain washing as reported by Mason and Buswell (1950). The quantity of silicates was comparatively low in the winter and summer, which might be attributed to its utilization in the formation of diatom walls.

SUMMARY AND CONCLUSION

The study of abiotic factors of Kalika pond (Dhar M.P.) revealed that the studied lentic habitat is perennial, shallow and alkaline water body. The abiotic factors viz. temperature, transparency, pH, dissolved oxygen, carbonates, bicarbonates, alkalinity, calcium, chloride, nitrate, phosphate and silicate showed marked seasonal variation in the present habitat, However free carbon-di-oxide could not observed from the surface water of the same. Their quantitative study revealed that water quality of this habitat is quite good for aquaculture and pisciculture.

ACKNOWLEDGEMENTS

Authors are thankful to Dr. K.S. Rao, Professor and Head, School of Studies in Zoology, Vikram University, Ujjain for some fruitful suggestion and constant encouragement.

REFERENCES

APHA (1985). Standard methods for the examination of water and waste water (16th ed.) APHA, AWWA and WPCF. Washington D.C

Bohra, O.P. and Bhargava, S.C. (1977). Abiotic factors, chlorophyll pigments and primary production in two lakes of Jodhpur. Geobios. **4**(5); 215-216.

Dave, R.K. and Shinde, D. (1986). Morphometry of the Kalika pond at Dhar (M.P.) J. Sci. Res. **8** (2&3) 111-115.

Ganapati, S.V. (1960). Ecology of tropical waters, Proc. Symp. Algol. ICAR, New Delhi **1**: 214-218.

Ganapati, S.V. (1962). A five year investigation of Almati reservoir Part 1-4. Environ. Hlth. **6** : 1-39.

Gonzalves, E. and Joshi, D.B. (1946). Freshwater algae near Bombay : The Seasonal succession of the algae in a tank at Bandra. J.Bomb. Nat. Hist. Soc. **46**(1) : 29-31.

Gupta, A.K. and Mehrotra,R.S. (1986). Studies on seasonal variations in pH and dissolved oxygen content in Sanhit Sarovar, Kurukshetra. Geobios. **13**(6) : 276-278.

Hannan,H.E (1979). Chemical modification in reservoir regulated streams. In the Ecology of Regulated streams. Ward, J.W. and Standford, J.A. Edition Plenum Corporation Publication : 75-94.

Hussainy, S.U. (1965). Environ. Hlth. **7**(1) : 24-31.

Hutchinson, G.E. (1957). A treatise on limnology. vol.1 Geography, physics and chemistry. NewYork, John Wiley & Sons, NewYork : 1015.

Jana, B.B. and Sarkar, H.L. (1971). The limnology of Swetganga. A thermal spring of Barkeswar, West Bengal, India Hydrobiol. **37**(1) : 33

Kaushik, S., Agarkar, M.S. and Saxena, D.N. (1991), Water quality and periodicity of phytoplanktonic algae in

Chambal Tal, Gwalior, Madhya Pradesh, Bionature, **11**(2) : 87-94.

Mason, W.P. and Buswell, A.M. (1950), Examination of water. John Wiley and Sons Inc., NewYork : 652.

Melzer, A. and Exler,D. (1982), Nitrate and reductase activities in aquatic macrophytes. Studies in aquatic vascular plants. J.J. Symoens, P. Compere, Royal Bot. Soc. Belg. Brusel : 128-135.

Michael, R.G. (1969), Studies on zooplankton in tropical fish pond, India. Hydrobiol, **32** (1&2) : 47-68.

Munawar, M. (1970), Limnological studies on fresh water ponds of Hyderabad, India. Hydrobiologia **31** : 101-128.

Nair, M.S.R., Seth, O.N. and Mishra, K.D. (1988), Limnological studies on a village pond, India (Vidisha). I Physico-chemical characteristics. Proc. Nat. Symp. Past, Present and Future of Bhopal Lakes : 195-201.

Nayak, T.R., Iyer, S.J. and Jaish, S.C. (1982), Seasonal variation of rotifers and certain Physico-chemical factors of Matyatal, Panna (M.P.), India. Comp. Physiol. Ecol. **7**(3) : 159-169.

Prakash, M.M. (1994), Seasonal variation in the physico-chemical characteristics of a perennial pond of Jhabua, M.P., J. Aqu. Biol. Fish **1**(1) : 43-47.

Prakash, M.M. (1996), Correlation Among the Physico-chemical characteristics of pond water (Mehta pond, Jhabua). In "Assessment of water Pollution" (S.R. Mishra ed.) PP. 441-451. APH Publishing Corporation, New Delhi, India.

Qudri, M.Y., Showakat, A., Naqash, G., Mustafa, S. and Yousuf, A.R., (1981), Limnology of two streams of Kashmir. J. India Inst. Sci. **63**(C) : 137-141.

Rajput, A. and Bhadoria, A.K.S. (1996), The Study in Hanumantal Lake of Jabalpur. In "Assessment of water Pollution" (S.R. Mishra ed.) pp. 342-348. APH Publishing Corporation, New Delhi, India.

Rao, K.S., Diwan, A.P. and Shrivastava, P. (1977), Limnological studies on a tropical lake, physical factors effecting Lophopodella carteri (Ectoprocta : phylactolaemata). Bioresearch (1-2) : 67-70.

Schwoerbel, J. and Tillmanns, G.C. (1977), Nitratonfnahme aur dam Wasser and Nitrate-duktase-Aktivital bei Fontinalis antipyretica. L. in Hell-Dunkd-Weschsel. Arch. Hydrobiol. Suppl., **48**: 412-423.

Shastree, N.K., Islam, M.S., Pathak, S. and Afshan, M. (1991), Studies on the physico-chemical dimensions of the Lentic Hydrosphere of Ravindra Sarovar (Gaya). In "Curr. Trends in limnology" (N.K. Shastree ed.) pp. 133-152, Narendra Pub. House, Delhi 1.

Shrivastava, P., Rao, K.S. and Diwan, A.P. (1977), Observation on the phenomenon of Eutrophication with relation to changing eulitoral planktonic productivity. Indian J.Zool., **5**(1) 37-39.

Singh,V.P. (1960), Phytoplankton ecology of the inland waters of Uttar Pradesh Proc. Symp. Alogology, ICAR, New Delhi: 243-271.

Singh, R.K. (1986), Fluctuation in the composition of zooplankton population in relation to hydrobiological conditions of a reservoir, J. Hydrobiol, **2**(3): 37-42.

Singh, S.B. and Sahai, R. (1979), Study of some limnological features of 'Jalwania' pond of Gorakhpur. Proc. Nat. Acad. Sci. India, **49**(B)iv : 207-215.

Sinha, A.B. (1969). Investigations on the ecology of Ramgarh Lake Gorakhpur University, Gorakhpur.

Sreenivasan, A. (1972), Limnology of a seepage type impoundment Odathurai tank. J. Indian Fish Soc. India, **4**: 162-168.

Sreenivasan, A. (1974), Limnological features of a tropical impoundment, Bhavanisagar reservoir (Tamil Nadu), India Int. Rev. Hydrobiol, **59**(3) : 327-342.

Takahashi, M. and Saijo, Y. (1981), Nitrogen metabolism in lake Kizaki, Japan. Ammonia and Nitrate uptake by phytoplankton. Arch. Hydrobiol, **91**: 393-407.

Toetz, D.W. (1981), Effect of pH, phosphate and ammonia on the rate of uptake of nitrate and ammonia by freshwater phytoplankton-Hydrobiol, **76**: 23-26.

Venkateshwarlu, V. (1969), An ecological study of the Moisi river, Hyderabad, India, with special reference to water pollution II : Factors influencing the distribution of algae. Hydrobiol. **33**: 352-363.

Verma, S.R. and Shukla, G.R. (1970), The physico-chemical conditions of Kamla Nehru tank, Muzzaffarnagar (U.P.) in relation to biological productivity, Environ Hlth., **12**: 110-128.

Vyas, L.N. (1968), Studies on the phytoplankton ecology of Pichhola lake, Udipur. Proc. Symp. on Rec. Adv. in tropical ecol, Part-1 : 334-347.

Yadav, Y.S., Singh, R.K., Choudhary, M. and Kotchhr, V. (1987). Limnology and productivity of Dighari Beed, Assam. Tropical Ecol., **28**: 137-146.

Zafar, A.R. (1964), On the ecology of algae in certain Fish ponds of Hyderabad, India. I physico-chemical complexes, Hydrobiol. **23**: 179-195.

Zafar, A.R. (1966), Limnology of the Husain Sagar lake, Hyderabad, India. Phykos., **5**: 115-126.

14

Colour Removal from Effluents of Paper Industries

Anita Nair & A.K. Shenvi

INTRODUCTION

The effluents of Pulp and Paper industry are coloured mainly because of lignin. The characteristics of effluents generated from such industries are represented in Table 1. Colour removal from waste streams is a difficult problem confronting this industry. Effluents from bleaching process are highly coloured and of significant volume typically 75-100 m^3 per ton of Pulp. The decolorisation and detoxification of bleachery effluents have become serious problems of industry in anticipation of increasingly stringent regulations specifying allowable toxicity levels and further, regulation of colour level of the Pulp mill effluents. Out of the total discharge of colour, more than 80% originates from bleach plant effluent. The colour load discharged in waste water from Pulp and Paper industry in India is recorded in Table 2. Pulp bleaching is necessary to brighten the pulp and the bleaching operation sequence depends on the type of Pulp bleached and the end product produced. The bleaching process, essentially is a delignification reaction. For chemical Pulps e.g. sulphate, sulfite and neutral sulfite, initial removal of lignin is achieved by cooking with alkalies, sulfides or sulfites. It is then followed by oxidative multistage bleaching using bleaching agents like chlorine (C), chlorine dioxide (D), hypochlorite (H) etc. and caustic extraction (E). The degree of brightness obtained is of course directly related to the degree of lignin removal obtained prior to oxidative bleaching. The application of any of the bleaching agent is

TABLE 1 : CHARACTERISTICS OF EFFLUENTS GENERATED FROM PULP AND PAPER INDUSTRIES

Particulars	S.S.	Kgs./Tonne of Paper B.O.D.	C.O.D.
Large Paper Mills (Bamboo and Hard Wood Based)	100-150	35-50	150-200
Agro Waste Based Small Paper Mills	90-240	85-270	500-1100
News Print Mills	100	45-50	135-140
Small Paper Mills Based on Waste Paper	50-80	10-40	50-90

Source : Data taken from "Environmental Pollution Due to Pulp and Paper Industry". IPPTA Seminar 1989 Silver Jubliee International Seminar & Workshop. Sept. 1989.

TABLE 2 : COLOUR LOAD DISCHARGED IN WASTE WATERS FROM PULP AND PAPER INDUSTRY IN INDIA

Particulars	Colour Kg./Tonne of Paper
Kraft Pulping	25-150
Kraft Bleaching	100-150
NSSC Pulping Recovery	100-125
Sulphite Pulping Recovey	15-100
Sulphite Bleaching	25-150

always followed by alkali Extraction (E). After each stage in multistage bleaching, a washing cycle is necessary. Therefore, water reuse is vital since 3-5 stage bleaching operation are quite common. Greater brightness can be achieved if smaller amounts of bleaching agents are applied in successive stages with washing between each stage, the resultant increase in water usage (as well as capital cost and fibre loss) is a limiting factor. With the development of chlorine dioxide as a bleaching agent, many modern bleach plant use only five stages to produce white and strong pulp.

Bleaching operation is the major source of waste water from the kraft pulp and paper mill, is with black liquor recovery furnace. The combined effluent from the bleaching section of a pulp and paper mill is highly coloured (Jones, 1973 and Halperm, 1975). In bleached kraft mills 60-70% of the total colour is typically derived from one effluent stream, the first caustic extract after chlorination. The colour discharged from various sections of large kraft paper mills is indicated in Table 3. The colouring body in the waste is organic in nature and is comprised of wood extractives, lignin and its degradation products formed by the action of chlorine or chlorine dioxide on lignin (Black and Willams, 1961). Lignin in the waste is present both as hydrophilically dispersed as well as low molecular wt. colourless soluble fraction. The later one is amenable to oxidation and former fraction is easily coagulated and flocculated and precipitated. The SS, BOD and COD loads from Agro residue based mills are reported to be 2.2, 3.4, and 2.7 times the corresponding load from that of large integrated paper mills. A number of treatment alternatives for Agro residue based and waste paper based mills have been prescribed.

TABLE 3 : COLOUR DISCHARGED FROM VARIOUS SECTIONS OF LARGE KRAFT PAPER MILLS

Source	Colour Pt-Co Unit
Washing and Screening	2,000
Chlorination	280
Caustic Extraction	20,000
Recovery Section Drain	200
Paper Machine	10
Mill Composite	1,000

A variety of methods for removing colour from effluents have been fairly effective in some situations. These include absorption by activated charcoal or synthetic resins on exchange, reverse osmosis, ultrafiltration, membrance technology, precipitation using alum and massive lime doses. The cost of these techniques is still fairly high and this restricts their use to specific local situations. The massive lime treatment as well as modified lime treatment of bleachery waste have been put to actual practice. The method however, has certain limitations viz., high pH and residual lemonade colour of the treated effluent, massive lime requirement etc. Therefore, still good deal of research is going on in this field to develop alternative methods. In view of this, an attempt is also being made in the laboratory to develop an economically and technically feasible treatment process to remove colour and toxic substances from the bleachery waste. Therefore, an effort was made to continuously treat coloured waste water from a bleached kraft mill in an upflow clarifier unit.

EXPERIMENTAL

The effluent from a bleached kraft Pulp and Paper mill was used for investigation. The waste water was stored in underground cement concrete tanks and the waste characteristics were regularly monitored. Commercial alum and clay were used as coagulants in this study. At first, jar tests were conducted to determine optimum dose and optimum clay dose along with alum. Sludge volume concentration and sludge volume indices of various sludges formed in jars were determined. Colour of the waste water and treated effluent was measured in accordance with the procedure (Carpenter and Berger, 1973).

The continuous operation of colour removal was studied in upflow clariflocculator of sludge blanket filtration type. Prior to the entry into the clarifier, waste was mixed with 2% alum and 2% clay solutions in calculated amounts. The dosage were controlled accurately by means of a constant head tank.

In order to evaluate the performance of the clariflocculator, a few parameters viz., colour removal, COD reduction, escape of suspended solids were taken into consideration. The waste was admitted to the clariflocculator at different rates ranging from 0.70 to 2.9 LPM equivalent surface loading rates being 7500-35,000 litres/sq metre/day. Initially the clariflocculator was run at a slower rate, which enabled the formation of sludge blanket. The sludge blanket thus formed, was able to trap the flocs coming at higher rate. Clariflocculator was run until flocs started escaping heavily over the outlet weir with the treated effluent i.e. until sludge blanket was at highest possible level. Samples were collected at the treated effluent outlet for each flow rate and analysed for various parameters. Flocculator was run using separately alum as a coagulant and to different day doses along with the optimum alum dose.

RESULT AND DISCUSSION

The effluent from kraft Pulp and Paper mill was first characterised. Table 4 records the characteristics of the untreated as well as treated waste. Initial colour of the effluent was 3500 units on pt -co scale. Optimum alum and clay requirements were determined by jar test. Maximum percentage of colour removal (94-95%) was taken as a criterion to fix 500mg/l of alum as an optimum dose. Addition of clay, does not in any way take part

TABLE 4 : CHARACTERISTICS OF UNTREATED AND TREATED EFFLUENTS

Parameter	Unit	Untreated effluent	Effluent Treated With		
			Alum 500 mg/l	Alum 500 mg/l & clay 500 mg/l	Alum 500 mg/l & clay 750 mg/l
Total solids	mg/l	1350	1296	1208	1177
*Suspended Solids	mg/l	–	360	41	33
pH	–	8.4	7.5	7.5	7.5
Alkalinity	mg/l	180	–	–	–
Acidity (CO_2)	mg/l	–	42	40	40
Sulphates	mg/l	310	464	464	464
** BOD_5	mg/l	511	422	422	422
COD	mg/l	1002	564	408	364
Colour	Co-Pt unit	3500	172	82	73
Residual Alum (as Al_2O_3)	mg/l	–	4.2	4.2	4.2

* **Suspended solids of untreated waste are negligible, since settled waste water used throughout the investigation.**

** **BOD_5 values are reported for jar test expts conducted on the day sample was collected. All other parameters are for the range of flow rates in the clari-flocculator.**

in colour removal by alum. Pelleting of the flocus and sludge takes place by mixing clay to the effluent alum mixture. Clay acts like a weighing agent for the light flocs. Therefore optimum clay dose was fixed on the basis of the minimum sludge volume concentration and sludge volume index. By such criteria, the optimum clay dose for the waste was 500 mg/l. The clay addition was necessitated by the fact that sludge volume concentration (SVC_{30}) was reduced from 25-30% (with alum alone) to 10% and sludge volume index was lowered from 450 (with alum alone) to 105. Based on these results, the colour of the waste was continuously removed in a clariflocculator using alum with and without clay. Two different doses of clay were used, optimum dose of 500 mg/l and 750 mg/l.

In an upflow clarifier, the quality of the treated effluents depends on the nature and size of the initially formed sludge blanket. The quality of the sludge blanket depends on the initial characteristics of the influent mixture (i.e. with or without clay) as well as the flow rate at which influent is admitted into the clarifier.

Sludge blanket from alum with 750 mg/l clay was more stable than the one from alum with 500 mg/l of clay.

The overall physical and chemical parameters for colourless effluent obtained by treating waste with optimum alum and alum with clay have been included in Table 4. To ascertain the operation parameters of the model colour, COD and SS removal were measured at different flow rates for 3 different treatments i.e. alum without any clay, alum with 500 mg/l clay and alum with 750 mg/l clay.

A high degree of colour removal was achieved in the laboratory flocculator. Residual colour of treated effluents at various flow rates have been recorded in Table 5. Colour removal of the order of 95-96% was achieved with alum and clay and 90% colour was removed when alum was used without any clay.

CONCLUSION

It is clear from results and discussion that clay assisted alum coagulation of the colouring matter can circumvent the problems encountered by massive lime treatment without additional cost of the treatment. The clarifier is a single unit in which flocculation, settling, pelletation, filtration and

TABLE 5 : RESIDUAL COLOUR IN TREATED EFFLUENT AT VARIOUS FLOW RATES. INITIAL COLOUR CONCENTRATION - 3500 CO-PT UNITS

Flow rate in the Flocculator (LPM)	Colour Alum only	Co-Pt Units Alum & 500 mg/l clay	Alum & 750 mg/l clay
0.60 (7200)	165	70	65
0.90 (9400)	180	–	–
1.20 (13252)	180	–	–
1.60 (17658)	180	80	80
1.80 (19296)	–	85	100
2.00 (21580)	–	95	105
2.20 (22692)	185	–	–
2.50 (24262)	190	100	110

absorption of soluble organics take place suggesting that it is a complete unit. COD of the treated effluent is well within the limits prescribed by ISI code. It is necessary to carry out detailed pilot plant studies (including the cost studies) before recommending its applicability on a large scale. Further it is better to look into possible disposal modes for 8-10% of the sludge of the total volume of the effluent treated.

ACKNOWLEDGEMENT

The authors are grateful to Management of Dandeli Education Society and to the Principal and Staff of B.N. Degree College, Dandeli (N.K.) for encouragement and support facilities.

REFERENCES

Black, A.P and Williams, D.G., (1961). Electrophonetic studies of coagulation for Removal of Organic Colour : Inc AWWA, **53** (5), 589-604.

Carpenter, W.L. and Berger, HF (1973). 'Development of improved procedure for measurement of Mill Effluent receiving water colour'. Proceed of the 28th Ind. waste conference, Purdue University, 234-251.

Halperm, M.G., (1975). Pulp Mill Processes-Pulping, Bleaching, Recycling, NDC London : England, p. 246.

Jones, H.R. (1973). "Pollution Control in the Pulp and Paper industry" NDC, London England, p. 246.

15

A Report on the Pre-pollution Status of the Vasishta Godavari Estuary, East Coast of India

A.G.R. Sai Sastry

INTRODUCTION

India has about 7000 km^2 of estuarine area spread over major riverine estuaries (Hooghly-Matlah system of West Bengal, Mahanadi of Orissa, Godavari and Krishna of Andhra Pradesh, Vellar-Coleroon complex of Tamil Nadu and those of Narmada and Tapti of Gujarat), a number of minor estuaries (like those of Mandovi and Zuari of Goa) and coastal brackishwater impoundments (Chilka lake of Orissa, Pulicat lake of Andhra Pradesh and Vembanad lake of Kerala). Each estuary has unique characteristics of its own depending upon its geographical location, quantum of land drainage it carries, nature of sewage, industrial effluents and farm drainage let into it, extent of tidal incursion and current systems operating in it.

The Godavari estuary is the largest monsoon-fed system of the subcontinent with the mighty river draining an area of 315, 980 km^2 of the Deccan plateau. It has an average discharge of about 45 million cusecs during the annual flood period. The river divides into two principal distributaries, Gautami to the east and Vasishta to the west, at Dowlaiswaram (close to Rajahmundry in Andhra Pradesh), the former opening into Bay of Bengal at Kottapalem (16°35'N 7 82°41'E) and Bhairavapalem (16°41'N & 82°23'E) and the latter at Antervedi (16°18'N & 81°42'E) and Vadalarevu (16°22'N & 81°96'E). A pioneering study of the Gautami Godavari estuary was undertaken by the scientists of Andhra University under the aegis of Indian

Council of Agricultural Research (ICAR) during 1958-63. Lack of any information relating to the Vasishta Godavari estuary till then and the threat of pollution with oil accompanying commercial exploitation of oil and gas wells located along its banks and off its mouth by oil and Natural Gas Commission (ONGC) from 1985-86 has prompted us to undertake a comprehensive survey of the system in 1981-83 to establish its pre-pollution status. In order to evaluate the impact of a pollutant on the ecosystem, it is necessary to monitor the physical and chemical characteristics of the system, the distribution and abundance of species and the normal variations of the characteristics over at least an annual cycle before, during and after the addition of the pollutant (Ganapati, 1975). As such, the results of the present survey are expected to serve as base-line data to assess the extent of damage caused to the estuary after the onset of pollution with oil. Our experience elsewhere has shown that pollution of estuarine/marine ecosystems with oil spills and effluents from oil-based industries results in a sea change of its water quality and large scale destruction of its plankton (Gabriel *et al*, 1975; Fernandez *et al*, 1977). The present paper deals with a brief overview of the results of the study.

The Godavari estuarine system experiences heavy annual flood under the impact of southwest monsoon. Depending upon the salinity of the water, it shows four distinct phases in an year. They are

(i) Annual flood phase coinciding with monsoon and incessant flow of freshwater as land drainage (July-September),
(ii) Early post-monsoon phase or recovery phase showing highly fluctuating salinity (October-December),
(iii) Late post-monsoon phase of stable salinity (January-March), and
(iv) Pre-monsoon phase or drought phase with total marine domination and consistently high salinity (April-June).

MATERIALS AND METHODS

The study includes sampling at the lower, middle and upper reaches of the estuary, at 1 km, 13 km and 28 km from its mouth at Antervedi. The periodicity of sampling was at weekly (middle reaches) and fortnightly (upper and iower reaches) intervals during 1981-82 and fortnightly/monthly intervals during 1982-83. A total of 59 and 21 sets of samples were studied from the middle and upper and lower reaches respectively to understand the spatial and temporal fluctuations of selected physico-chemical factors and plankton. The water samples were collected from the surface, close to the bottom (9.5-10.0 m) and two intermediate depths (4 m and 8 m) covering all the phases of the tide. Temperature, salinity, dissolved oxygen and inorganic nutrients - PO_4-P, NO_2-N and SiO_3-Si-of the water were measured using appropriate methods already described (Sai Sastry & Chandramohan, 1990a). Plankton samples were collected by oblique hauls using a nylo-bolt townet (143 μm and 0.5 m dia) fitted with a calibrated TSK flowmeter. Estimates of standing crop were made using volume displacement method and of the biomass (no./m^3) of each taxonomic group through screening the entire sample for larger forms and through aliquot analysis for others. Since the river was in spate during the annual flood with literally liquid mud flowing down, it was not possible to continue sampling during the said phase.

OBSERVATIONS / RESULTS

The ranges of physico-chemical factors observed were summed up in Table 1. Basic information collected on plankton was shown in Tables 2 and 3.

DISCUSSION

Hydrography

(a) Tidal amplitude and impact : The estuary experiences semi-diurnal tides of the adjoining Bay of Bengal with an amplitude of 0.5m and 1.3m during neap and spring tides. The onset of the tide in the estuary was delayed by 20-45 min. Depending upon the distance from the mouth and the resistance encountered by the penetrating seawater from the fresh/brackishwater rolling down from above.

(b) Temperature : The role of temperature in tropical estuaries is only marginal, because of their geographical location close to the equator. However, in the Vasishta

TABLE 1 : RANGES OF PHYSICO-CHEMICAL FACTORS IN THE VASISHTA GODAVARI ESTUARY (OCT. 1981 TO JUNE 1983)

Factor		Lower reaches	Middle reaches	Upper reaches
Temperature (°C)	S	24.2 – 32.2	21.5 – 34.0	24.7 – 33.5
	B	24.2 – 31.9	21.0 – 32.5	24.6 – 33.3
K		1.7 – 11.3	0.8 – 17.0	0.8 – 6.8
Salinity (%)	S	21.0 – 34.4	0.5 – 30.9	0.5 – 20.0
	B	23.0 – 34.8	0.5 – 32.0	0.7 – 20.4
Dissolved	S	3.6 – 6.4	2.6 – 8.1	3.7 – 7.2
oxygen (ml/l)	B	3.3 – 5.5	2.2 – 6.3	3.8 – 6.7
DO Saturation (%)	S	77 – 131	56 – 183	71 – 131
	B	72 – 118	51 – 121	82 – 123
Nutrients (μ gm at/l)	S	ND – 1.0	ND – 10.0	ND – 3.3
NO_2 – N	B	ND – 1.2	ND – 3.8	ND – 2.5
PO_4 – P	S B	0.1	0.1 – 0.2	ND – 0.1
S_iO_3 – Si	S	13.0 – 50.0	10.0 – 358.9	13.5 – 416.7
	B	10.0 – 41.6	10.0 – 208.3	12.5 – 454.5

S = Surface; B = Bottom ND = Not Detected

TABLE 2 : PLANKTONOLOGICAL CHARACTERS OF THE VASISHTA GODAVARI ESTUARY

Character	Lower reaches	Middle reaches	Upper reaches
Standing crop (ml/m^3)	0.5 – 15.9	0.1 – 26.7	1.0 – 19.8
Total Nos./m^3	2080–519330	910–2708590	31930–1315970
Adults (%)	22.9 – 97.3	9.6 – 85.6	14.3 – 83.6
Larvae (%)	2.6 – 77.1	14.4 – 90.3	16.4 – 85.7
Coelenterates (%)	0.0 – 0.2	0.0 – 0.07	0.0 – 0.004
Copepods (%)	18.7 – 96.6	6.0 – 81.6	4.6 – 77.1
Other Crustaceans (%)	0.0 – 0.09	0.0 – 0.1	0.0 – 0.01
Invertebrate larvae (%)	2.6 – 77.0	14.4 – 90.3	16.3 – 85.3
Chaetognaths (%)	0.01 – 1.1	0.0 – 3.5	0.0 – 1.2
Pelagic tunicates (%)	0.04 – 17.2	0.0 – 32.0	0.0 – 4.3
Fish eggs & Larvae (%)	0.0 – 6.2	0.0 – 0.9	0.0 – 0.4
Miscellaneous groups (%)	0.0 – 0.9	0.0 – 17.6	0.0 – 78.9

TABLE 3 : ZOOPLANKTERS OF THE VASISHTA GODAVARI ESTUARY

Taxonomic group	Total no. of species	Representative forms
Coelenterates	28	Species of *Aglaura, Phialidium, Blackfordia, Bougainvillia, Acromitus, Pleurobrachia, Beroe* etc.
Copepods	33	Species of *Paracalanus, Acrocalanus, Eucalanus, Centropages, Labidocera, Acartia, Pontella, Pseudodiaptomus, Oithona, Macrosetella* etc.
Other Crustaceans (Mysids, Amphipods, Isopods, Decapods etc.)	18	Species of *Rhopalophthalmus, Neomysis, Potamomysis, Siriella, Caprella, Paracaprella, Cirolana, Urothoe, Lucifer* etc.
Chaetognaths	3	Species of *Sagitta*
Pelagic tunicates	2	Species of *Oikopleura*
Miscellaneous groups (Rotifers, Polychaetes, Pteropods etc.)	16	Species of *Brachionus, Keratella, Filinia, Asplanchna, Hexarthra* etc., *Tomopteris, Travisiopsis Creseis*
Invertebrate Larvae	28	Cerianthids Polynoids, Phyllodocides, Syllids, Nepthyids, Spionids, Mitrarids and Magelonids Naupliids, Protozoeae (of *Acetes* & *Lucifer*), Zoeae of crabs, Megalopas, Alimas, Mysids Veligers of Gastropods and Bivalves Bipinnarids, Plutids & Auricularids Actinotrochs, Brachiopod larvae Cyphonautids Pilidia

Godavari estuary, the temperature of the water was basically dependent upon neritic penetration of cool seawater in the lower reaches, upon insolation, addition of warm farm drainage, stratification of the water column, and tidal currents in operation in the middle reaches and upon insolation alone in the upper (depth 1.75m). In the small hours of winter days, the surface water turned cooler than the subsurface water due to fall of temperature of the atmospheric air.

(c) Attenuation coefficient (K): The turbidity of the water column measured as Attenuation coefficient was consistently high during the flood phase and early recovery phase obviously because of the silt contained in. During the drought phase, churning of the bottom mud at the confluence by high winds from the south kept the water of the lower reaches highly turbid in contrast to clear water higher up.

(d) Salinity: The influence of penetration of seawater was felt inside the estuary up to a distance of 40 km from the mouth, though the impact declined gradually with increasing distance. The average salinity of the water column rose from 22.5 to 34.5‰, 0.5 to 31.4 ‰ and 0.6 to 20.2‰ in the lower, middle and upper reaches respectively from the recovery to the drought phase. A tongue of seawater was found at the bottom in the fore estuary even during the height of annual flood. Seawater was found locked up throughout the year in a 15 m-deep canyon in the middle reaches with tidal currents moving up and down over it without disturbing it because of the density gradient prevalent.

(e) Dissolved oxygen: The concentration of dissolved oxygen (column average) varied from 3.5 to 5.8 ml/l, 2,7 to 6.9 ml/l and 3.8 to 6.6 ml/l in the lower, middle and upper reaches respectively. The percentage of saturation ranged from 56 to 183. Total absence of 'anoxic' conditions indicated absence of pollution with degradable organic wastes, industrial effluents and additives like oils that consume the oxygen contained in or interfere with diffusion of atmospheric oxygen into water to the detriment of living organisms inhabiting the estuary.

(f) Inorganic nutrients: Of the three inorganic nutrients studied, the concentration of phosphate-phosphorus was found to be much lower than in the Gautami Godavari estuary, probably because of absence of mangrove swamps that act as storehouses for phosphate, keeping it locked in mud ready to release into water whenever its level goes down. The nitrite-rich farm drainage flowing into the estuary on and off from the rich agricultural hinterland kept its levels generally high until it was utilized fully to trigger production of blooms of phytoplankton.

The spatial and temporal variations of physico-chemical factors and the diel and tidal fluctuations of the water quality of the Vasishta Godavari estuary were already described in detail (Sai Sastry & Chandramohan, 1990a & b).

Phytoplankton

Though the present study primarily aims at analysing zooplankton of the estuary, such of those phytoplankters that clogged the townet and contributed to the bulk of the standing crop of plankton were also studied. Of the 30 genera identified, *Chaetoceros, Coscinodiscus* and *Thalassiothrix* in the lower reaches, *Bacillaria, Chaetoceros, Rhizosolenia, Thalassiosira* and *Thalassiothrix* in the middle reaches and *Campylodiscus* and *Coscinodiscus* in the upper reaches often formed blooms. The bloom of *Gymnodinium splendens* (dinoflagellate) whenever it appeared excluded other phytoplankters altogether and reduced most of the zooplankters drastically. Seasonal fluctuations in phytoplankton production showed primary peaks in October/November, November-June and March/April in the lower, middle and upper reaches of the estuary respectively. Excessive production of phytoplankton and its gross underutilization by zooplankton was noticed in the lower reaches.

Zooplankton

The zooplankton collected during the survey had about 100 species of adults of various taxonomic groups and 28 types of invertebrate larvae, a distinct measure of qualitative richness by any standard. While biodiversity of the plankton and planktonic coelenterates of the Vasishta Godavari estuary were described in detail (Sai Sastry & Chandramohan, 1994 & 1989), a comprehensive report on the pre-pollution status of zooplankton is in press (Sai Sastry & Chandramohan, 1995). The seasonal fluctuations of zooplankton in the estuary showed basically a trimodal pattern in the lower and middle reaches and bimodal pattern in the upper. Primary peaks were observed in February-May, November and March-May in the lower, middle and upper reaches respectively. The impact of tides and day and night cycles on different groups of zooplankton was studied over a 24-hour period once in a month in the middle reaches from November 1981 to May 1982 (Sai Sastry, 1987). Some of the important findings brought out by the study of zooplankton of the estuary are as follows:

1. *Aglaura hemistoma, Phialidium hemisphaericum, P. globosum, Blackfordia virginica* and *Bougainvillia fulva* dominated the planktonic coelenterates of the estuary and were largely confined to the lower and middle reaches.
2. Of the 33 species of copepods identified, 26 were calanoids. *Paracalanus parvus, P. aculeatus, Acartia erythraea, A. spinicauda, Oithona rigida* and *Euterpina acutifrons* were present throughout the estuary. *Acrocalanus gibber, Eucalanus crassus, Centropages tenuiremis, C. dorsispinatus, Labidocera acuta, L. pectinata* and *pontella securifer* were confined only to the lower reaches, while *Clausocalanus arcuicornis, Pseudodiaptomus binghami* and *Acartia Clausi* were seen only in the upper. Copepods formed the single largest group in the zooplankton of the estuary (40.6 to 60.3%) and were abundant in the middle reaches (50.2%). The major peaks in their seasonal distribution were between February to May (S = 29.634.2 ‰), October/November (S = 5.6-19.6 ‰) and March to May (S = 13.9-20.0 ‰) at the lower, middle and upper reaches respectively.
3. Among the other crustaceans, *Rhopalopthalmus kempi* (mysid) and *Lucifer hanseni* (sergestid shrimp) were dominant and present throughout the non-flood period in the lower and middle reaches.
4. Among the invertebrate larvae, those of crustaceans and molluscs dominated the estuary in the lower and middle and middle and upper reaches respectively. Polychaete larvae, because of their marine origin, were confined only to the lower reaches. A general rise in the numbers of larvae was noticed in late post-monsoon and early pre-monsoon months (February and May) in the lower and upper reaches and in the middle of early recovery, stable and drought phases (November, February and May) in the middle. A keen competition between the larvae of crustaceans and molluscs, expressed as decline of one with an increase of the other, was noticed throughout the estuary.
5. Among chaetognaths, *Sagitta bedoti* was dominant (96%) and was present throughout the estuary, though

the body size of the individuals and density of its population declined with a decrease in salinity towards the head. It was observed to breed more than once in an year as shown by juveniles appearing in batches at least 3-4 times during the non-flood period.

6. Majority of the rotifers recorded in the estuary were of thecate type and were primarily of limnetic origin. They were mostly seen in the upper and middle reaches during the early recovery phase.
7. Swarms of pteropod, *Creseis acicula* were seen drifting into the estuary from neritic waters on and off.

CONCLUSION

Absence of anoxic conditions in the estuary, qualitative and quantitative abundance of plankton, frequent appearance of phytoplankton blooms and ability of the ecosystem to utilize inorganic nutrients fully indicates that Vasishta Godavari estuary was a healthy ecosystem, without pollutants of any type affecting its water quality and biota as things stood in 1981-83.

With the onset of annual flood, all the plankters of the Vasishta Godavari estuary are flushed down into the adjoining sea and the fast moving silt-laden water turns the estuary into a hostile habitat unfit for any plankter. When the flood recedes, the estuary is repopulated through gradual recruitment of euryhaline neritic forms from the adjoining sea, moving into it with seawater. As such, any factor that affects the neritic environment and its populations has profound influence on the qualitative and quantitative richness of the estuarine flora and fauna.

Fifty onland wells yielding natural gas (free and associated with oil) and sixteen offshore wells yielding oil have been identified in the Krishna-Godavari (KG) basin so far. Of these 13 are on banks of Vasishta Godavari estuary. The major Ravva formation (along with G-1, G-2, G-15, G-23 and G-29), comparable to Bombay High in richness is in the neritic waters midway between the openings of Gautami and Vasishta Godâvari estuaries into the sea. The gas drawn out from the onshore wells (approx. 1.7 million m^3 per day at present) is carried through a system of criss-crossing pipelines to the bulk storage depots for supply to gas-based industries for generation of electricity, production of fertilizers and a few more units including a paper mill. The oil taken out is being sent to the refinery at Visakhapatnam either through road or through coastal liners. The potential of the gas/oil wells of the area can be gauged from the Pasarlapudi blowout (on 8th January 1995) in Gas well No. 19 located on the banks of Vasishta Godavari's shorter branch (Vainateyam), it being the biggest ever encountered by ONGC in its 50-years history. There are already reports of oil coming out of the onshore wells along with gas seeping into the farmlands surrounding the wells and from there into the estuary through farm drainage. We can reasonably presume that the commercial exploitation of the oil wells of Ravva basin going on since early 1990s must have had its impact on the seawater by now. Hence, the need of the hour is a post-pollution survey to assess the extent of damage caused to the ecosystem.

ACKNOWLEDGEMENTS

The author acknowledges with thanks the receipts of financial assistance from the University Grants Commission (UGC), New Delhi under Faculty Improvement Programme during the tenure of which the present study was carried out.

REFERENCES

Fernandez George, A., Daniel, A. and Nicket, P. (1977). Oil spills in the marine environment at Madras and its impact on Neuston and Plankton during 1975-76. *Proc. Symp. on Warmwater Zooplankton* NIO/UNESCO Spl. Publ. 264-273.

Gabriel, P.L., Dias, N.S. and Nelson-Smith, A. (1975). Temporal changes in the plankton of an industrialized estuary. *Estuar. Coast. Mar. Sci. 3(2)* : 141-151.

Ganapati, P.N. (1975). Estuarine Pollution. *Bull. Dept. Mar. Sci. Univ. Cochin. VII*, **1**:1-9.

Sai Sastry, A.G.R. (1987). Studies on the hydrography and plankton of a tropical tidal estuary (The Vasishta Godavari) on the east coast of India. Ph.D. Thesis approved by Andhra University.

Sai Sastry, A.G.R. and Chandramohan, P. (1989). Planktonic coelenterates of Vasishta Godavari estuary, east coast of

India. *Indian J. Mar. Sci.* Vol. **18** : 160-164.

________ & ________ (1990a). Physico-chemial characteristics of Vasishta Godavari estuary, east coast of India; Pre-pollution Status. *Ibid.* Vol. **19** : 42-46

________ & ________ (1990b). Diel and tidal fluctuations in the water quality of Vasishta Godavari estuary. *Ibid.* Vol. **19** : 133-136.

________ & ________ (1994). Biodiversity in the plankton of Vasishta Godavari estuary, east coast of India. *J. Aqua. Biol.* Vol. *9 (1&2)* : 11-15.

________ & ________ (1995). Zooplakton of Vasishta Godavari estuary, a pre-pollution status survey. *J. Mar. Biol. Ass. India.* Vol. *37* (In press).

16

Trophic Structure of Plankton Community in Some Typical Wetlands of Kashmir, India

Ashok K. Pandit

INTRODUCTION

The living organisms are now well known as the causative agents for polluting the environment. Symptoms of eutrophication (increase in organic production due to nutrient enrichment) that seem to concern people most are the unwanted phytoplankton blooms. Zooplankton, the immediate consumers of phytoplankton, form the next higher trophic level in the food-chain after phytoplankton. And in fact a close relationship exists between the two life-form classes. Changes in the physico-chemical properties of water, therefore, alter both the composition and abundance of phytoplankton or seston and thus affect the consumers of successive trophic levels.

The interest in plankton productivity studies has rapidly increased during last two decades or so, to a large extent due to the impetus given to it by the I.B.P. programme (1964-1974). However, a majority of experimental field studies pertain to population dynamics while only little seems to have been done in regard to the trophic relations of plankton (Enright, 1969; Gliwicz, 1969; Gliwicz and Hillbright-Ilkowska, 1972; Haney, 1971, 73; Hillbright-Ilkowska, 1971; Wright, 1958; Gulati, 1975; Anderson, 1970; Anderson and Green, 1975; Pandit, 1980) and their use as indicators of trophic status of a lake or a waterbody (Jarnefelt, 1952, 56, 63; Rawson, 1956; Davis, 1961, 64; Jumppanen, 1976; Kaul *et al.* 1978; Kaul and Pandit, 1979; Pandit, 1980; Pandit and Kaul, 1981). But so far as the socioeconomically important and highly productive wetland ecosystems are concerned, there is practically no literature available both at the country level and the state level except for a few short report (Kaul *et al.* 1978; Pandit and Kaul, 1981; Kaul and Pandit, 1982). The present study, therefore, aims at the elucidation of various ecological aspects of micro-cosm community vis-a-vis the trophic structure.

DESCRIPTION OF THE STUDY AREA

The study was carried in five different wetlands namely, Nowgam, Malgam, Haigam, Mirgund and Hokarsar, all situated in the floodplains of River Jhelum in the northwest of the city of Srinagar (1580-1600 m altitude, 34°5'-34°20' N, 74°5'-74°9' E). The area, depth range and the nature of various wetland sites (Fig.1) are given as under :-

Wetlands	Area (km^2)	Depth Range (m)	Type of Wetland
1. Nowgam	3.05	0–0.80	Temporary and Shallow
2. Mirgund	4.29	0.05-1.05	Temporary and Deep
3. Malgam	4.16	0.22-0.40	Permanent and Shallow
4. Haigam	14.00	0.15-0.80	Permanent and Shallow
5. Hokarsar	5.00	0.66-1.22	Permanent and Deep

Out of five wetlands Hokarsar and Haigam are maintained by the state Government as the two biggest bird sanctuaries in Kashmir as they provide overwintering resorts to about 0.2 million waterfowl (ducks, geese and rails) migrating from the palaearctic region, extending from north Europe to central Asia, and breeding grounds to a host of other birds in summer. The wetlands are also important as rich sources of table fish (*cyprinus carpio communis* and C.C. *specularis*), and of fodder and other economically important plants (Pandit and Fotedar, 1982; Pandit, 1988).

The drainage basins of these wetlands are mainly comprised of level arable land under paddy cultivation, the water from the varied streams and channels irrigating these fields often accumulates in them. They, for some part of the year, are rainfed and serve as sinks during floods. Haigam and Hokarsar have regular inlets and outlets for preventing them from drying out in winter. Potable water oozing out from a large number of springs scattered all along the margins and within

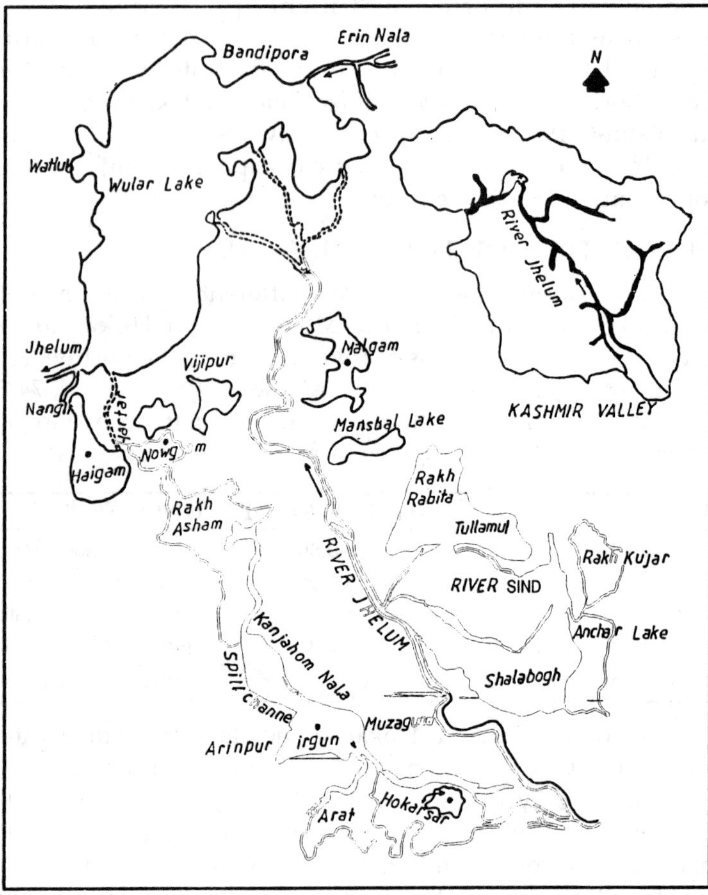

FIG. 1

the wetland basin is an additional but permanent source of water for Malgam (Kaul and Pandit, 1980).

Textural composition of wetland sediments reveals them to be silt-loam type, with silt, clay and fine sand being the major constituents of the sediment system. The distinctive feature of Malgam is the peaty nature of its soil. The climate and vegetation of the wetland areas have been described elsewhere by Kaul (1980, 82), Kaul *et al.* (1978, 80) and Pandit (1980, 84, 88, 91).

GENERAL TROPHIC STATUS OF THE WETLANDS

The physico-chemical characteristics of water in different wetlands is given in Table 1. The study reveals that the water temperature is closely connected with the air temperature, and in the shallow water it also determines the oxygen concentration with which it shows inverse relation. The chemical characteristics of the alkaline and hard water in wetlands show pronounced annual variations which are determined by the autotrophic bio-activity especially for nutrients and CO_2. Since wetlands are shallow and contain relatively small volume of water compared with the area of contact with sediments, the oxygen consumption is fast and as such lower oxygen values, falling even below 3 mg/l during summer, in comparison to lakes are justifiable (Kaul and Trisal, 1984). The water analysis of various wetlands reveal pH, ranging between 7.2-8.6, fall towards alkaline side. Of the divalent ionic components, Ca and Mg are the most abundant. The usual cation progression is Ca>Mg>Na>K. In Kashmir wetlands the concentration of divalent cations, in general, is very high while bicarbonates is the most dominant anion. According to OHLE (1934) the waters can be classified as "Calcium rich" and, therefore, "hard water type" of MOYLE (1945). The concentration of free CO_2 is generally low, the dissolved inorganic carbon necessary for photosynthesis being obtained from bicarbonates. Magnesium depicts an irregular trend though slightly higher values are recorded in spring and fall. This is also true for Na while K shows a definite seasonal trend, the values increasing during autumn probably due to higher decomposition rates at that time (Pandit, 1980; Kaul and Trisal, 1984).

Despite large supplies of nutrients from the catchment area, low values of dissolved inorganic phosphate phosphorus (DIPP) seem to prevail in wetlands of Kashmir (Table 1). Total phosphate phosphorus (TPP) is present in ample quantities, the depletion of DIPP from these waters being associated with photosynthetically induced precipitation, besides its utilization by the autotrophs. The high pH values, associated with high temperature and high alkalinity, are favourable conditions for precipitation of $CaCO_3$ due to rapid carbon assimilation from dissolved bicarbonates. Under such conditions, phosphate ion is reported to coprecipitate with carbonate (Otsuki and Wetzel,

TABLE 1 : AVERAGE VALUES (A) AND ANNUAL RANGES (R) OF VARIOUS PHYSICO-CHEMICAL PARAMETERS OF WATER IN FIVE DIFFERENT WETLANDS OF KASHMIR

Parameter		Malgam	Haigam	Nowgam	Mirgund	Hokarsar
Temperature	A	17.50	17.17	19.80	17.17	16.41
(°C)	R	6.50–28.60	6.00–29.06	12.42–27.31	7.00–28.85	6.14–27.85
Conductivity	A	569.84	480.00	526.70	533.75	505.75
(μ mhos/cm)	R	471 – 873	309 -- 701	346 – 829	304 – 1012	357 – 638
pH	A	7.57	7.80	7.79	8.02	7.75
	R	7.70–8.80	7.38–8.21	7.48–8.04	7.18–8.70	7.47–8.52
Alkalinity	A	192.42	122.84	135.60	172.50	168.34
(mg/l)	R	105.30–299.30	68.12–169.50	93.00–212.42	95.71–249.28	98.71–249.00
Chlorides	A	8.57	8.50	10.47	8.27	8.21
(mg/l)	R	6.80–10.70	7.36–11.12	8.57–13.14	6.42–10.85	6.71–9.50
DIPP (μg/l)	A	33.58	38.25	50.72	43.17	39.25
	R	11.70–64.00	22.62--78.62	26.71–77.14	25.14–81.85	14.42–54.57
TPP (μg/l)	A	254.84	236.17	227.70	274.34	210.08
	R	110.40–673.30	65.25–552.36	112.28–657.00	118.56–511.42	112.00–568.30
DOPP (μg/l)	A	253.50	215.84	227.70	234.25	205.25
	R	80.90–619.50	50.62–518.36	67.57–583.71	79.42–429.57	80.42–415.41
Dis. Oxygen	A	6.18	6.20	5.78	5.84	5.81
(mg/l)	R	4.48–8.42	4.54–9.20	4.55–9.00	3.14–7.80	4.50–8.33
NO_2–N	A	8.91	10.25	10.50	8.17	9.08
(μg/l)	R	0–19.50	0 – 23.50	1.70–20.42	0.50–31.85	0 – 16.57

NO_3–N	A	288.50	234.41	289.50	300.25	247.33
(µg/l)	R	43.50–644.00	35.35–621.86	41.42–556.00	46.00–624.14	73.14–618.00
NH_3–N	A	25.33	33.50	39.70	26.08	36.58
(µg/l)	R	23.30–54.20	12.86–77.87	10.85–61.14	17.14–39.00	16.57–69.87
Silicates	A	12.50	10.91	9.10	7.84	9.84
(mg/l)	R	6.00–18.00	8.00–16.00	0 – 14.00	6.00–10.00	9.00–12.00
Na (mg/l)	A	12.00	10.91	9.09	7.54	9.64
	R	6.35–18.05	8.52–16.24	5.59–14.54	7.43–10.13	9.32–12.55
K (mg/l)	A	2.02	1.60	2.75	2.97	1.93
	R	1.01–4.62	0.71–2.55	1.03–6.16	1.38–40.92	1.27–2.73
Ca (mg/l)	A	54.58	45.50	49.60	50.50	42.75
	R	35.90–85.80	26.00–68.65	35.14–73.28	33.57–84.14	29.46–50.42
Mg (mg/l)	A	16.78	13.00	13.90	14.41	13.41
	R	12.40–24.80	6.50–17.86	8.56–22.71	11.00–23.28	11.00–17.42

A = Average **R = Range**

1972; Kaul *et al*, 1978). The coprecipitation of phosphate ion with $CaCO_3$ being the main mechanism of phosphate-phosphorus removal from the waters is borne out by the fact that relatively higher amounts of unreactive PO_4-P occur at the time of maximum $CaCO_3$ precipitation. This unreactive PO_4-P probably represents the coprecipitated portion of phosphorus (Kaul *et al.*, 1978).

NO_2-N and NH_3-N are present in low quantities in comparison to NO_3-N which is the main nitrogen source in these aquatic bodies. The increase in NO_3-N concentration towards winter and early spring (Pandit, 1980) may be attributed to the combined result of nitrification at mud-water interface and the vigorous stirring of wetland waters which eliminates any limiting effects of oxygen diffusion (Handoo, 1978). In polluted waters, high quantities of NO_2-N and NH_3-N (total upto more than 50 mg 1^{-1}) have been determined. The increase in NH_3-N during spring and fall in a consequence of decomposition of organic matter (Keeney, 1973; Kaul and Trisal, 1984), as NH_4-N is an end product of nitrate metabolism.

In general, higher values of alkalinity and DIPP and lower values of oxygen are recorded in wetlands as compared to lakes (Kaul *et al.*, 1978). The nutrient level of water, however shows a great variation in different wetlands depending upon the geological substratum, soil type, vegetation, water depth, and human activities etc.

METHODS

(a) Sampling

Water samples were taken, usually from the trophogenic layer in a water column, at each of the sampling stations differing in water depth, vegetation, food supply and other characteristics on monthly basis. The samples were thoroughly mixed in equal proportions to form a representative sample of 10 litres which was sieved through a plankton net of bolting silk (No. 20). The sieved residue was transferred to a bottle and preserved in modified Lugol's solution (Pandit, 1980) and 4% formalin, the latter improving preservation and inhibiting any bacterial contamination and growth (Utermohl, 1958; Nauwerck, 1963; Vollenweider, 1969; Anderson and Green, 1975).

(b) Counting

Quantitative estimation of plankton was done with the help of Sedgewick-Rafter cell of 1 ml capacity. The technique devised by Utermohl (1958) was employed. The zooplankton and unicellular algae were counted as individuals while in filamentous Cyanophyceae 100 μm lengths of the filament were taken as the equivalents. Similarly the filamentous chlorophyceae were recorded as cells while in the colonial forms like *Gomphosphaera, Microcystis, Pandorina* and *Volvox* etc. the counting unit was a colony (Jumppanen, 1976).

(c) Plankton Biomass

The plankton biomass was estimated according to Lohmann (1908). Since the volume and fresh weight of plankton very much correspond to each other the mean density being 1.01-1.03 according to Hutchinson (1967)- the biomass is expressed as micrograms per litre ($=10^6 \times \mu^3/1$) or mg/m^3 assuming $1\ \mu^3=10^{-9}$ mg (Spodniewskae, 1971; Jumppanen, 1976). The average cell volume of each taxon was computed after giving it an appropriate geometrical shape after Vollenweider (1969), Naulapaa (1966), Nauwerck (1963) and Jumppanen (1976).

(d) Laboratory Experiments

Two methods were employed to assess the feeding habits of zooplankton, the first consisting in the observations on feeding under the microscope and the second in the analysis of gut contents. To confirm the field observations, a series of laboratory experiments were conducted with the zooplankton, higher animals and other varied food materials and thus the food preferences of various species were ascertained. Trophic chains of plankton were drawn by linking together the results obtained from the various sets of experiments.

RESULTS

(a) Phytoplankton

(1) Species Composition

The species diversity, especially in terms of species numbers, is exceptionally great in all the wetlands under study. Majority of the species occur regularly and none is strictly limited to any one biotope in particular, a fact corroborating the general eutrophy of all the wetland sites. A total of 190 species and sub-species are identified in the phytoplankton of which 20 belong to cyanophyceae, 72 to chlorophyceae, 60 to Bacillariophyceae, 3 each to Xanthophyceae, Chrysophyceae and Dinophyceae and 29 to Eugleninae. The maximum number of species is, however, recorded in Nowgam (189), followed by Haigam (183), Mirgund (179), Malgam (177), and decreasing to minimum in Hokarsar (156).

Although the mean number of species in wetlands under study, in general, is high, yet only a comparatively small number of species contribute appreciably to the total density and biomass, presumably owing to the state of dynamic balance of plankton community; the species competing for the same materials in a fairly uniform environment. In such a process some species profit more than others and hence dominate the scene. The various dominant forms under different classes are :

Cyanophyceae : *Anabaena, Nostoc, Oscilliatoria, Lyngbya, Microcystis, Gomphosphaera* and *Merismopedia.*

Chlorophyceae : *Chlamydomonas, Chlorella, Tetraedron, Crucigenia, Scenedesmus, Pediastrum, Akistrodesmus, Ulothrix, Chaetophora, Coleochaete, Oedogonium, Zygnema, Spïrogyra, Mougetia, Cosmarium, Closterrium, Arthrodesmus* and *Staurastrum.*

Bacillariophyceae : *Navicula, Cymbella, Pinnularia, Gomphonema, Amphora, Synedra, Diatoma, Fragilaria, Epithemia, Anomoneis, Coconeis, Gyrosigma* and *Surirella.*

Eugleninae : *Euglena, Trachelomonas, Phacus, Lepocynclis, Astatia* and *Strombomonas.*

(2) Density and Biomass

Seasonal variation in the density and biomass of phytoplankton revealed two peaks of growth, one in spring and the other in autumn. Vigorous mass outbursts at peak productive stages often result in the formation of algal or phytoplankton blooms. During summer, the phytoplankton making the primary production, show only modest appearance both in regard to density and biomass against their minimum touched in winter or ice-cover period.

Of the sites studied, the mean biomass corresponding to average density was greatest in Malgam (3,927,600 ind. and 188.8 mg m^{-3}), followed by Haigam (3,450,000 ind. and 186.1 mg m^{-3}), Mirgund (3,513,500 ind. and 173.5 mg m^{-3}), Nowgam (2,706,600 ind. and 140.8 mg m^{-3}) and decreasing to minimum in Hokarsar (402,300 ind. and 18.9 mg m^{-3}). However, in all the wetlands under study, Chlorophyceae (predominantly the filamentous forms of green algae) appear to contribute a decisive proportion (67.70 – 87.92%) of the primary production by phytoplankton throughout the year. Bacillariophyceae, Cyanophyceae and Eugleninae, though moderately represented, are also responsible for the remaining part of phytoplankton density and biomass. The insignificant contributions of poorly represented groups like Xanthophyceae (*Vaucheria* sp. and *Botryodiopsis* sp.), Chrysophyceae (*Dinobryon sertularia, D. sociale* and *Chrysamoeba radians*) and Dinophyceae (*Ceratium hirundinella, Cymnodinium* sp. and *Peridinium* sp.) in the overall production of phytoplankton is a noteworthy feature (Tables 2-3, Fig. 2). This variation in the relative contribution of different groups together with a simultaneous increase in total biomass are clear signs of fast eutrophication.

(b) Zooplankton

(1) Species Composition

The 141 species of zooplankton, encountered in all the wetlands, are spread over various taxonomic groups with the following break up :

(i) Protozoa – 61 (42 rhizopods + 17 ciliates + 2 zooflagellates)
(ii) Rotifera – 29
(iii) Crustacea – 51 (35 cladocerans + 13 copepods + 3 ostracods)

The number of species in various wetlands is given as under :

	Nowgam	Malgam	Haigam	Mirgund	Hokarsar
Protozoa	48	47	51	39	44
Rotifera	21	19	25	17	18
Crustacea	40	43	40	41	38
Total	**109**	**109**	**116**	**97**	**100**

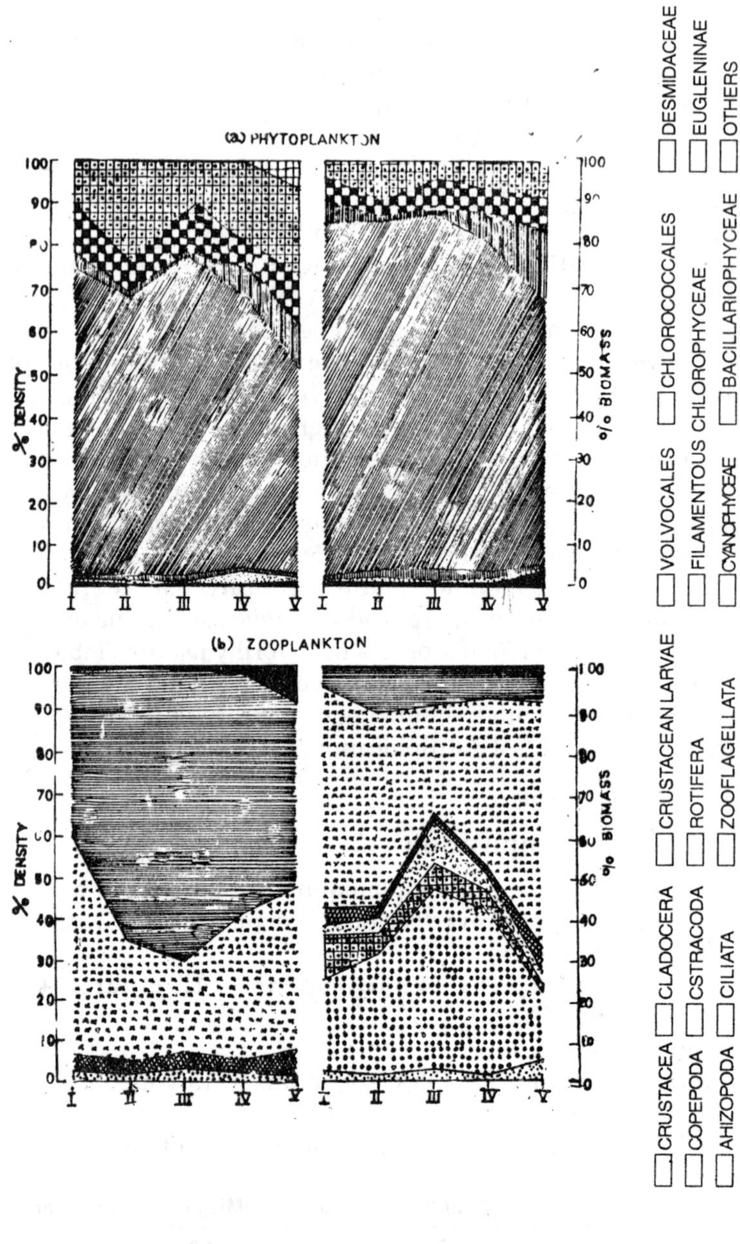

I = NOWGAM II = MALGAM III = HAIGAM IV = MIRGUND V = HOKARSAR

TABLE 2 : AVERAGE DENSITY (X 10^4 IND. m^{-3}) OF PHYTOPLANKTON IN FIVE WETLANDS

Taxonomic groups	Nowgam	Malgam	Haigam	Mirgund	Hokarsar
Cyanophyceae	31.25 (11.6)	27.57 (7.0)	34.41 (9.8)	24.75 (7.0)	4.57 (11.3)
Chlorophyceae	201.80 (74.5)	267.91 (68.2)	267.66 (77.6)	242.10 (68.9)	20.79 (51.7)
Volvocales	2.80 (1.0)	0.77 (0.2)	0.82 (0.2)	0.89 (0.3)	0.34 (0.9)
Chlorococales	2.46 (0.9)	5.95 (1.5)	0.98 (0.3)	11.98 (3.4)	0.60 (1.5)
Desmidiales	1.29 (0.5)	4.08 (1.0)	7.11 (2.1)	3.13 (0.9)	0.12 (0.3)
Filamentous Chlorophyceae	195.26 (72.1)	257.11 (65.5)	258.75 (75.0)	226.10 (64.3)	19.73 (49.0)
Bacillariophyceae	26.51 (9.8)	93.45 (23.8)	39.19 (11.4)	63.76 (18.2)	8.18 (20.3)
Xanthophyceae	0.39 (0.1)	0.10 (0.03)	0.04 (0.01)	0.20 (0.06)	2.41 (6.0)
Chrysophyceae	0.06 (0.02)	0.02 (0.01)	0.06 (0.02)	0.08 (0.02)	0.20 (0.05)
Dinophyceae	0.08 (0.03)	0.14 (0.04)	0.08 (0.03)	0.05 (0.01)	0.01 (0.03)
Eugleninae	10.57 (3.9)	3.56 (0.9)	3.56 (1.0)	20.40 (5.8)	4.07 (10.1)
Total	**270.66**	**392.75**	**345.00**	**351.35**	**40.23**

Relative densities (%) are given in parenthesis.

TABLE 3 : COMPARISON OF FIVE DIFFERENT WETLANDS ON THE BASIS OF MEAN BIOMASS (MG m^{-3}) AND THE PERCENTAGE BIOMASS CONTRIBUTION (IN PARENTHESES) OF PHYTOPLANKTON

Taxonomic groups	Nowgam		Malgam		Haigam		Mirgund		Hokarsar	
Cyanophyceae	7.5	(5.4)	6.4	(3.4)	12.4	(6.6)	8.9	(5.1)	1.6	(8.5)
Chlorophyceae	120.7	(85.8)	162.3	(85.9)	163.7	(87.9)	140.2	(80.3)	12.8	(67.7)
Volvocales	1.7	(1.2)	1.9	(1.0)	1.4	(0.8)	2.2	(1.3)	0.8	(4.2)
Chlorococales	0.7	(0.5)	2.5	(1.3)	0.5	(0.3)	1.8	(1.2)	0.1	(0.4)
Desmidiales	1.1	(0.8)	3.6	(1.9)	6.3	(3.4)	2.2	(1.6)	0.1	(0.6)
Filamentous Chlorophyceae	117.2	(83.3)	154.3	(81.7)	155.5	(83.4)	131.4	(76.2)	11.8	(62.5)
Bacillariophyceae	4.9	(3.5)	17.2	(9.1)	7.2	(3.9)	11.7	(6.7)	1.5	(7.9)
Xanthophyceae	0.2	(0.1)	+		+		0.1	(0.1)	0.1	(0.3)
Chrysophyceae	+		+		+		+		+	
Dinophyceae	0.5	(0.3)	0.6	(0.3)	0.5	(0.3)	0.1	(0.1)	+	
Eugleninae	7.0	(4.9)	2.3	(1.3)	2.3	(1.3)	13.5	(7.7)	2.9	(15.5)
Total	**140.8**		**188.8**		**186.1**		**173.5**		**18.9**	

The most important and frequent genera among the zooplankton are :

Protozoa : *Amoeba, Arcella, Centropyxis, Pontigulasia, Difflugia, Nebela, Pareuglypha, Euglypha* and *Actinosphaerium* among Rhizopoda and *Paramecium, Porodon, Colpoda, Colpidium, Euplotes* and *Holophrya* among ciliata.

Rotifera : *Keratella, Brachionus, Cephalodella, Monostyla, Lecane, Synchaeta, Asplanchna, Trichocera, Filina* and *Mytilina.*

Crustacea : *Daphnia, Ceriodaphnia, Simocephalus, Scapholeberis, Bosmina, Graptoleberis, Alona, Alonella* and *Chydorus* among cladocera; *Cyclops, Eucyclops, Mesocyclops, Bryocamptus* and *Canthocamptus* and *Cypris* and *Eucypris* among Ostracoda.

(2) Density and Biomass

In general, two peaks of zooplankton are observed, one in late-spring – early summer and the other in autumn. Though the density maxima obtained in fall session surpass the peak recorded in the early session of the year except for Hokarsar, yet the biomass peak at all the wetland sites turned to be somewhat sub-dominant under the greatly surpassing biomass peak registered in late spring–early summer possibly because of the greater biomass contributions of crustaceans in the warm water period. The low populations of zooplankton, in general, during ice cover period and crustaceans particularly in August 1976 (a period of heavy floods) together with the rareness of *Diaptomus* sp. throughout the sampling period are among the most obvious differences.

Among the sites studied, the maximum zooplankton standing crop was recorded in Mirgund (63.81 mg m^{-3}), followed by Haigam (38.34 mg m^{-3}), Malgam (31.17 mg m^{-3}), Nowgam (27.81 mg m^{-3}) and decreasing to minimum values in Hokarsar (7.64 mg m^{-3}), the zooplankton volume at the first site being apparently eight times that of Hokarsar. The average density values followed a trend similar to that of mean biomass excepting zooplankton density at Malgam surpassing that of Haigam (Tables 4 and 5, Fig. 2). In all the wetlands, however, the decisive proportions of the zooplankton standing crop is due to Crustacea and Protozoa, the contribution of Rotifera being

TABLE 4 : AVERAGE DENSITY ($X10^4$ INDIV. m^{-3}) OF ZOOPLANKTON IN FIVE WETLANDS

Taxonomic groups	Nowgam	Malgam	Haigam	Mirgund	Hokarsar
Protozoa	13.27 (93.6)	20.98 (95.5)	18.59 (92.9)	31.56 (95.0)	5.20 (92.6)
Rhizopoda	7.54 (53.2)	6.57 (29.9)	4.56 (22.8)	11.96 (36.0)	2.24 (39.9)
Ciliata	5.60 (39.5)	14.31 (65.1)	13.89 (69.4)	18.90 (56.9)	2.45 (43.6)
Zooflagellata	0.13 (0.9)	0.10 (0.5)	0.14 (0.7)	0.70 (2.1)	0.52 (9.1)
Rotifera	0.56 (3.9)	0.68 (3.1)	0.83 (4.1)	0.99 (3.0)	0.35 (6.2)
Crustacea	0.34 (2.4)	0.32 (1.4)	0.59 (3.0)	0.65 (2.0)	0.06 (1.1)
Crustacean larvae	0.04 (0.3)	0.03 (0.01)	0.07 (0.4)	0.04 (0.2)	0.02 (0.4)
Cladocera	0.12 (0.9)	0.19 (0.9)	0.34 (1.7)	0.38 (1.2)	0.03 (0.5)
Copepoda	0.17 (1.2)	0.08 (0.3)	0.13 (0.6)	0.18 (0.5)	0.01 (0.2)
Ostracoda	0.01 (+)	0.02 (0.1)	0.06 (0.3)	0.04 (0.1)	+
Total	**14.17**	**21.98**	**20.01**	**33.20**	**5.61**

Relative Densities are given in Parantheses

TABLE 5 : MEAN BIOMASS (MG m^{-3}) OF ZOOPLANKTON IN FIVE WETLANDS

Taxonomic groups	Nowgam	Malgam	Haigam	Mirgund	Hokarsar
Protozoa	16.11 (57.9)	17.89 (57.4)	13.31 (32.7)	30.54 (47.8)	5.15 (67.5)
Rhizopoda	14.69 (52.8)	14.30 (45.9)	9.80 (25.6)	25.67 (40.2)	4.49 (59.0)
Ciliata	1.40 (5.0)	3.58 (11.5)	3.49 (9.1)	4.80 (4.5)	0.61 (8.4)
Zooflagellata	0.02 (0.1)	0.01 (+)	0.02 (+)	0.07 (0.1)	0.05 (0.1)
Rotifera	1.14 (4.1)	0.84 (2.7)	1.04 (2.7)	1.22 (1.9)	0.43 (5.6)
Crustacea	10.56 (38.0)	12.44 (39.9)	23.99 (63.6)	32.05 (50.2)	2.05 (26.9)
Crustacean larvae	0.63 (2.2)	0.46 (1.5)	1.07 (2.8)	0.91 (1.4)	0.35 (4.6)
Cladocera	6.23 (22.4)	9.39 (30.1)	16.98 (44.3)	25.45 (39.9)	1.28 (16.7)
Copepoda	3.13 (11.3)	1.41 (4.5)	2.30 (6.0)	3.28 (5.1)	0.18 (2.4)
Ostracoda	0.57 (2.1)	1.18 (3.8)	3.64 (9.5)	2.41 (3.8)	0.24 (3.2)
Total	**27.81**	**31.17**	**38.34**	**63.81**	**7.63**

Relative Biomass (%) are given in Parentheses

comparatively low. This greater contribution of Crustacea and Protozoa is, in turn, due to Cladocera (16.72-44.28% of the total biomass) and Rhizopoda (25.56–59.00%) respectively.

(c) Planktonic Food-chains and Food Web

The food habits of various zooplankton species inhabiting the five wetlands of Kashmir are summarized elsewhere by Pandit (1980) and Pandit and Kaul (1981).

(A)
Gombusia offinis
Eucyclops sp. & cladocerans
Phytoplankton

(B)
Gombusia offinis
Cyclops scuifer
Ceratium sp.
Detritus

(C)
Cyclops bicolor
Daphnia sp.
Chlorella sp.
Decayed Ceratophyllum

(D)
Mesocyclops hyalinus
Gastropus stylifer
Polyarthra vulgaris
Cryptomonas sp.

(E)
Diaptomus sp.
Cyclops vemalis
Chydorus sp. Alona sp.
Chlorophyceae

(F)
Odonate nymph
Mesocyclops sp.
Brachionus sp.
Eupoltes sp.
Detritus

(G)
Gammarus sp.
Diaptomus sp.
Cyclops scutifer
Asplanchna priodonta
Keratella cochlearis
Cryptomonas sp.

(H)
Odonata
Lecane sp., Filina sp., Mytilina sp. and Colurella sp.
Actinospherium sp.
Pelomyxa sp.
Colpoda sp.
Microcystis sp.

(I)
Gambusia affinis
Diaptomus sp.
Cyclops sp.
Asplanchna sp.
Brachionus Calyciflorus
Arcella sp.
Diatoms

(J)
Cyclops bicuspidatus
Diaptomus sp.
Cyclops vernalis
Synchaeta sp.
Cephalodella sp.
Monostyla sp.
Amoeba sp.
Paramecium sp.
Chlamydomonas sp., Chlorella sp. & Bacteria
Detritus

FIG. 3 : A SERIES OF FOOD CHAINS IN KASHMIR WETLANDS BASED ON TROPHIC RELATIONS OF PLANKTON (PANDIT, 1980; PANDIT AND KAUL, 1982)

According to different trophic levels, in the food-chain, the zooplankton inhabiting the wetlands can be classified into (i) primary consumers which feed directly on plant material (algae and bacteria), (ii) secondary consumers which in turn graze upon the primary consumers, and (iii) tertiary consumers consuming secondary consumers. Even omnivores, acting as both primary and secondary consumers, are distinguishable.

The diversity of zooplankton communities coupled with the variety of potential foods, comprising phytoplankton, zooplankton, detritus, and bacteria suggest much complex and variable nature of trophic relations in the food-chains that lead to the building of a series of linear food-chains each consisting of a number of steps (Fig. 3) and the formulation of a scheme depicting the typically complex but generalized type of food web prevalent in these wetland sites (Fig. 4).

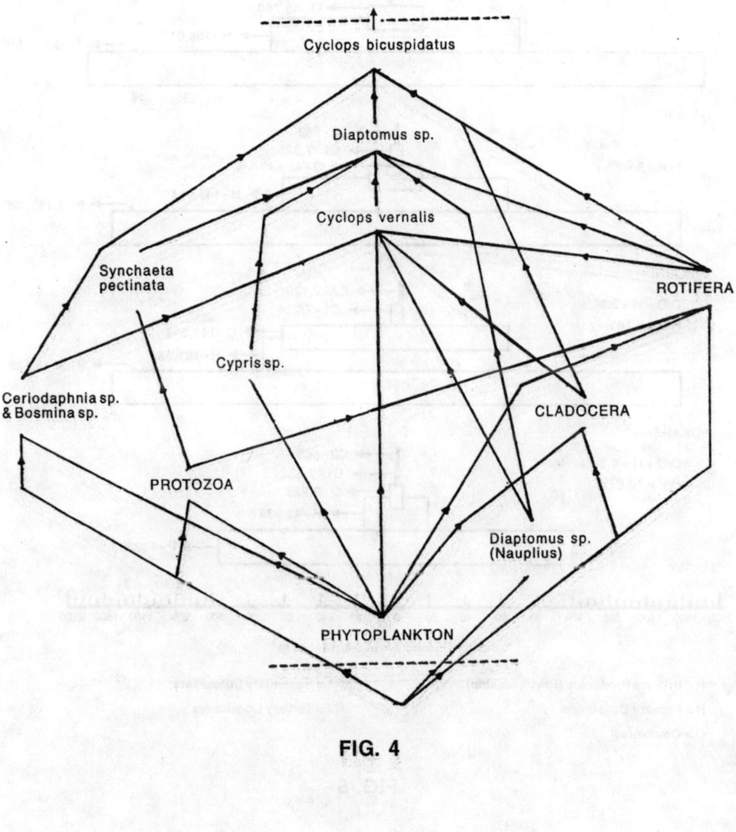

FIG. 4

The constituent links, over the entire length of trophic chains considered here, at the outset, suggest two main features, i.e.,

(i) that each link in the food-chain has a food source restricted to biomass than the density, and

(ii) that the accelerated eutrophication causes a shortening of food-chain since primary consumers largely switch from planktonic algae to bacteria and detritus.

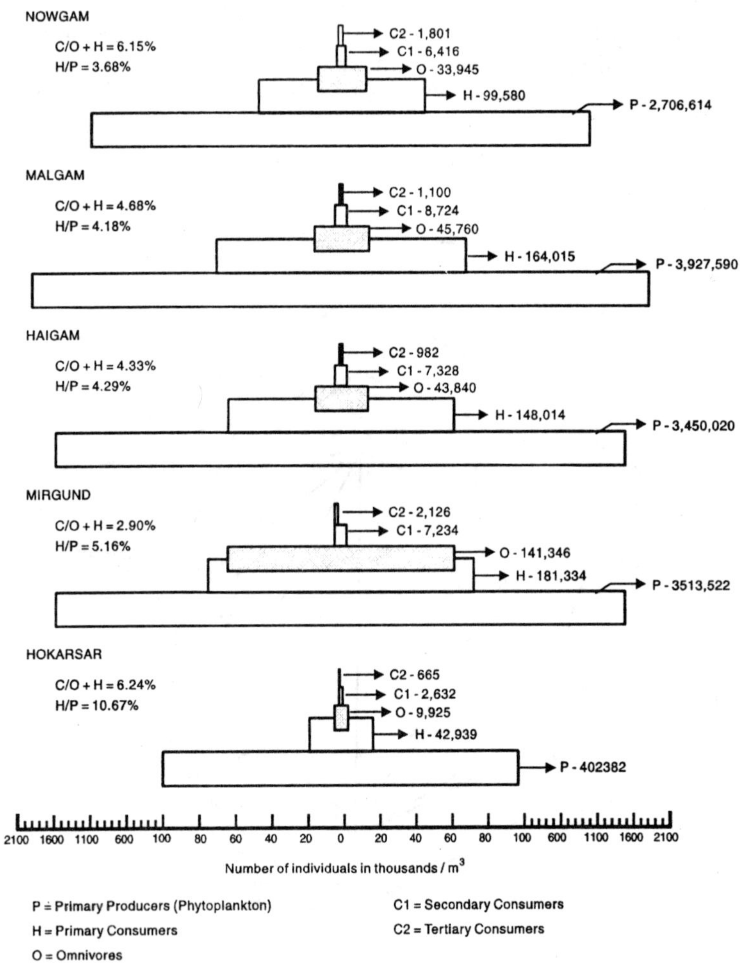

FIG. 5

(d) Trophic Structure of Plankton Community

The data collected under various trophic levels were finally utilized in setting out the quantitative trophic structure in the form of pyramids of number (Fig. 5) and biomass (Fig. 6). Despite certain limitations and errors in making the finer estimates (Pandit, 1980), the typical upright pyramid structure in all the cases was almost similar and according to expectations.

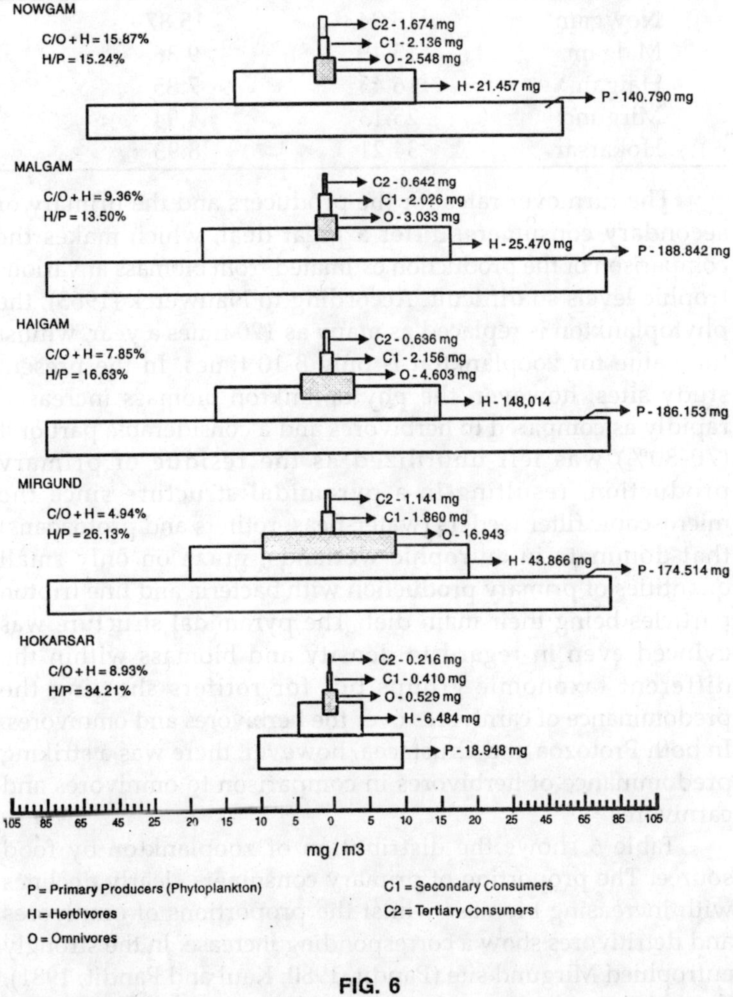

FIG. 6

The pyramids for the various sites, however, differed on quantitative basis. A comparison of biomass at various trophic levels in the pyramid gives an idea of the relative levels of production in different wetlands. Thus, the ratio of standing crop between trophic levels is :

Site	H/P (%)	C/O+H (%)
Nowgam	15.24	15.87
Malgam	13.49	9.36
Haigam	16.43	7.85
Mirgund	25.13	4.94
Hokarsar	34.21	8.93

The turn over rates for the producers and the primary or secondary consumers, differ a great deal, which makes the comparison of the production estimated from biomass at various trophic levels so difficult. According to Nauwerck (1963), the phytoplankton is replaced as many as 120 times a year, whilist the value for zooplankton is only 8-10 times. In the present study sites, however, the phytoplankton biomass increased rapidly as compared to herbivores and a considerable part of it (70-80%) was left unutilized as the residue of primary production, resulting in a pyramidal structure since the microscopic filter feeders (water-fleas, rotifers and protozoans) that dominate in eutrophic wetlands, graze on only small quantities of primary production with bacteria and fine tripton particles being their main diet. The pyramidal structure was evinced even in regard to density and biomass within the different taxonomic groups but for rotifers showing the predominance of carnivores over the herbivores and omnivores. In both Protozoa and Crustacea, however, there was a striking predominance of herbivores in comparison to omnivores and carnivores.

Table 6 shows the distribution of zooplankton by food source. The proportion of primary consumers clearly declines with increasing biomass whilst the proportions of omnivores and detritivores show a corresponding increase. In the strongly eutrophied Mirgund site (Pandit, 1980; Kaul and Pandit, 1981), the primary consumers comprised 68.74% by weight and 54.61% by density as compared to the least eutrophied

Hokarsar site where the corresponding values were as high as 84.68% and 76.46% respectively. A similar but less pronounced change, may be discerned in Malgam, Nowgam and Haigam sites also. Since the primary consumers mainly use the phytoplankton as food, it appears that only a part of net primary production is utilized by the zooplankton in strongly eutrophied biotopes. An important part, presumably in the sediments, is decomposed by bacteria and then utilized by detritivores.

DISCUSSION

The study reveals a gradual increase in the number of species as well as the primary production as a result of eutrophication. But further the conditions diverge from the normal and optimum values of the majority of the species, the smaller is the number of species in a community and more is the density of only a few species left over (Thienemann, 1939). Harsh limiting factors decrease the number of species and the diversity and also reduce the stability (Macarthur, 1955). A heavy increase in nutrients during final phases of eutrophication also lower diversity (Margalef, 1961) since some species profit more rapidly from such an increase than others and become dominant. In the zooplankton, a decrease in diversity occurs as a result of eutrophication, especially during the blooms of blue-green algae. This further leads credence to the Patten's (1962) "Species Diversity Principle" according to which pollution limits the number of species and produces dense populations of a few left over. And this explains the relatively low species number and high plankton production at Mirgund and Malgam as compared to Nowgam and Haigam sites.

According to Jumppanen (1976) the phytoplankton has proved to be a rather more sensitive indicator of slight eutrophication as compared with chemical properties of water for even if the changes in the nutrient content are very low, their combined influence may evoke an appreciable reaction in the phytoplankton.

With an increase in phytoplankton biomass, the proportion of green algae (predominated by filamentous forms and Chlorococcales), followed by blue-greens, some dominant diatoms and euglenoids increased whilst that of Xanthophyceae, Chrysophyceae, Dinophyceae, Desmidiales and

TABLE 6 : DISTRIBUTION OF ZOOPLANKTON BY FOOD HABITS

Wetlands	Herbivores (E) %		Omnivores and Detritivores (O) %		Carnivores (C1) %		Carnivores (C2) %	
	Density	Biomass	Density	Biomass	Density	Biomass	Density	Biomass
Nowgam	70.25	77.14	23.95	9.16	4.53	7.68	1.27	6.02
Maigam	74.62	81.71	20.91	9.73	3.97	6.50	0.50	2.06
Haigam	73.95	80.71	21.90	12.00	3.66	5.62	0.50	1.66
Mirgund	54.61	68.74	42.57	26.56	2.18	2.91	0.64	1.79
Hokarsar	76.96	84.68	17.67	6.92	4.69	5.37	1.18	2.83

TABLE 7 : RELATION BETWEEN PHYTO- AND ZOOPLANKTON IN KASHMIR WETLANDS

Wetlands	Biomass					Density	
	% Phytoplankton (P)	% Herbivorous Zooplankton (H)	P/H	Phyto-plankton (%)	Zoo-plankton (%)	Phyto-plankton (%)	Zoo-plankton (%)
Nowgam	86.78	13.22	6.56	83.50	16.50	95.02	4.98
Malgam	88.12	11.88	7.47	85.83	14.17	94.70	5.30
Haigam	85.74	14.26	6.01	82.92	17.08	94.52	5.48
Mirgund	79.91	20.09	4.00	73.23	26.77	91.37	8.63
Hokarsar	74.50	25.50	2.92	71.27	28.73	87.75	12.25

some diatoms (e.g., *Coconeis* sp., *Ceratoneis* sp., *Synedra acus*, *Tabellaria* sp., *Diatomella balfouriana*, *Astronella formosa*, *Cymatopleura solea*, *Cyclotella stelligera* and *Gyrosigma* sp.) besides *Sphaerocystis schroeteri*, typical of oligotrophic conditions, decreased, making the whole plankton association somewhat more eutrophic. These changes involving the major increase in the proportion of nutritionally inferior algae like Chlorophyceae over the decreased proportions of Chrysophyceae, Xanthophyceae, Dinophyceae and Desmidaceae, in addition to accelerating contributions of Cyanophyceae, Eugleninae and some diatoms and the simultaneous increase in the total biomass are clear signs of racing eutrophication (Thunmark, 1945; Jarnefelt, 1956; Rawson, 1956; Davis, 1964; Jumppanen, 1976; Pandit, 1980; Pandit and Kaul, 1981). These facts together with plankton quotient system (not included in this study–Pandit, 1980) lend support to the proposition of classification of wetlands into various types of eutrophic waterbodies. As such Mirgund shows vigorous eutrophication and appears to be a fully matured waterbody. A similar pronounced rise in the trophic status can also be discerned in Malgam, Haigam and Nowgam are moderately eutrophicated when compared with Hokarsar which is still passing through infant stages of eutrophication.

Eutrophication also affects the species composition, biomass and structure of zooplankton community. At least during the initial phases, the biotic factors seem to be much more effective than the chemical factors, the changes being largely determined by alterations in the phytoplankton or the trophic conditions. Chrysomonads, important as food, are replaced by green and blue-green algae, one of the important reasons for the replacement of indicators of oligotrophy like *Diaptomus* sp., *Cyclops bicuspidata*, *Sida crystallina*, *Pseudosida bidentata*, *Bosmina coregoni* and *Daphnia middendorffiana* by eutrophic types like Rotifera, Cladocera and Copepoda, the former species being encountered in small numbers only. The predominance of *Keratella* sp., *Brachionus* sp., *Cephalodella* sp., *Monostyla* sp., *Lecane* sp., *Asplanchna priodonta*, *Bosmina longirostris*, *Ceriodaphnia* sp., *Daphnia pulex*, *Scapholebris* sp., *Simocephalus* sp., *Graptolebris testudinaria*, *Chydorus* sp., *Alona* sp. and *Cyclops vernalis* besides the group Protozoa as a whole in comparison to above mentioned oligotrophic crustaceans particularly *Cyclops bicuspidata* and very rare *Diaptomus* sp. are clear signs of fast eutrophication (Jarnefelt, 1952, 56, 63; Rawson, 1956; Davis, 1961, 64; Pejler, 1964; Patalas and Patalas, 1966; Patalas, 1972; Anderson, 1974; Tundsi and Tundsi, 1974; Jumppanen, 1976; Kaul *et al.*, 1978; Kaul and Pandit, 1979; Pandit, 1980; Pandit and Kaul, 1981).

The rich crop of Protozoa, especially in Malgam and Mirgund, may be attributed to the higher amounts of organic matter providing the basic source of food. The higher percentage of rotifers may similarly be attributed to the availability of rich protozoan food.

The studies have revealed the increase in zooplankton biomass due to eutrophication, being highly favourable for the zooplankton at various wetland sites, resulting in a consequent increase in both number and biomass of zooplankton to the extent characteristic of eutrophic waterbodies (Shushkina, 1966; Patalas, 1968; Jumppanen, 1976; Widuto, 1977; Pandit, 1980). However, the vigorous eutrophication ultimately seems to limit the rate of increase in zooplankton biomass below that of primary production level. The proportion of phytoplankton biomass is high during the period of algal blooms and as such the changes in the biomass ratio (Table 7) can conveniently be used as another indicator of trophic status. According to Jumppanen (1976) the ratio between phytoplankton and zooplankton biomass was 0.27 in oligotrophic Haapovesi and 1.66 in very eutrophic Tuokkolanthi. The wetlands under the present study, however, show an exceedingly great ratio, the phytoplankton biomass being manifold (4-5 times) greater than that of zooplankton, a situation quite contrasting to the one obtained by Nauwerck (1963) according to whom the phytoplankton biomass never exceeds zooplankton biomass in the eutrophic Lake Erken. Since nutrient value of planktonic algae decreases in eutrophied biotopes, bacteria and detritus form a large proportion of omnivorous filter feeders.

A comparison of biomass at various trophic levels (Fig. 6) and the structure of biomass pyramid in wetlands with both cultural and natural eutrophication gives an idea of the relative levels of production. In the less eutrophic Hokarsar site, the proportion of phytoplankton biomass (74.5) is comparatively lower than that of highly eutrophic Malgam (88.12%) site and correspondingly the proportion of herbivorous zooplankton is greater (25.5%) at the former as compared to the latter site (11.88%). Nowgam and Haigam, being moderately eutrophicated, occupy the intermediate position. Such a study clearly indicates that during eutrophication the phytoplankton biomass grows more rapidly than that of herbivores. But in the highly eutrophicated Mirgund site such a relationship between the biomasses seems to be reversed as has also been reported by Jumppanen (1976). This, in all likelihood is due to rapid turn over of algal populations and the bacterial production, as a comparatively small phytoplankton biomass seems to be sufficient to maintain a large biomass of primary consumers in the less eutrophic biotopes. In the highly eutrophic wetlands, however, a considerable part of phytoplankton biomass remains unutilized. The role of planktonic algae in the zooplankton diet is further diminished by a notable decrease in the proportion of primary consumers in the zooplankton biomass, as is clearly indicated by the distribution of zooplankton by food source (Table 6). In the strongly eutrophied Mirgund site the primary consumers comprised 68.74% by weight and 54.61% by density as compared to least eutrophied Hokarsar where the corresponding values were 84.68% and 76.46% respectively. The unedible algae presumably sediments and is decomposed by bacteria and then utilized by detritivores. This may result in a sharp increase in decomposition and also shortening of food chain, since primary consumers largely switch from planktonic algae to bacteria and detritus. Thus, at critical times of the year, the food web of an extremely eutrophied wetland may largely consist of producers and decomposers as also opined by Jumppanen (1976).

Estimates of actual production and energy flow in lake ecosystems have been shown to indicate a notable increase in the phytoplankton production to herbivorous zooplankton with increasing eutrophication by Winberg (1972). A similar situation is also obtained in the wetlands as indicated by P/H ratio (Table 7). The lower consumption of planktonic algae may be attributed to the unfavourable structure of the plankton community (blue-greens being rejected as food by water-fleas) and the gradual disappearance of large herbivorous crustaceans. The ecological niche ordinarily harbouring such animals as a consequence remains unoccupied and may at times disturb the equilibrium of the ecosystem as also suggested by Jumppanen (1976), according to whom the larger planktonic herbivores compose a weak but rather important link in the food chain in the process of rapid eutrophication. The same very reason explains the numerical surge of large cladocerans (e.g., *Daphnia* sp.) being preceded by a short lived increase of small cladocerans like *Bosmina longirostris, Alona* sp. and *Chydorus* sp. and a marked increase in the number of *Cyclops vernalis*. Such a shift is presumably a predator avoidance mechanism helping the small crustaceans to escape predation by fish (*Gambusia affinis* and fry of *Cyprinus carpio*), odonate nymphs, *Notonectus undulata, Nepa* sp., *Dytiscus marginalis* and other predators as also corroborated by the studies of Brynildson and Kempinger (1973), Anderson (1974) and Richards *et al*., (1975). Brook (1969) observed a shift in dominance from large bodied planktonic herbivores (as a result of size dependant predation) being very much indicative of a higher degree of eutrophication. The mass occurrence of *Daphnia pulex*, tolerant to pollution, at few stations is due to the fact that predaceous animals and fish feeding on zooplankton usually avoid polluted biotopes (Jumppanen, 1976).

It is likely that predator-prey relationship, competition and permanence of waterbody are the most important factors influencing the distribution pattern of certain ecologically important species like *Cyclops vernalis, C. bicuspidata,* and *Diaptomus* sp. and the dimensions of food web. Since *C. bicuspidatus* preys heavily on *Diaptomus* sp. both seem to be positively associated and, therefore, both decline. Such a decline may also be attributed to their oligotrophic nature (Carl, 1940; Tundisi and Tundisi, 1976; Anderson and Green, 1976). The dominance of *Cyclops vernalis* in wetlands may further be due to little predation by *Diaptomus* sp. (Anderson, 1970). This further supports the viewpoint held by Anderson (1970, 1972) that *C. bicuspidatus* declines in presence of more abundant *C. vernalis*. Thus, *Diaptomus* sp. and *C. bicuspidatus* association again forms a weaker link in the food-chain of wetlands, the most important food links showing the possible flow of materials being depicted in Figure 7.

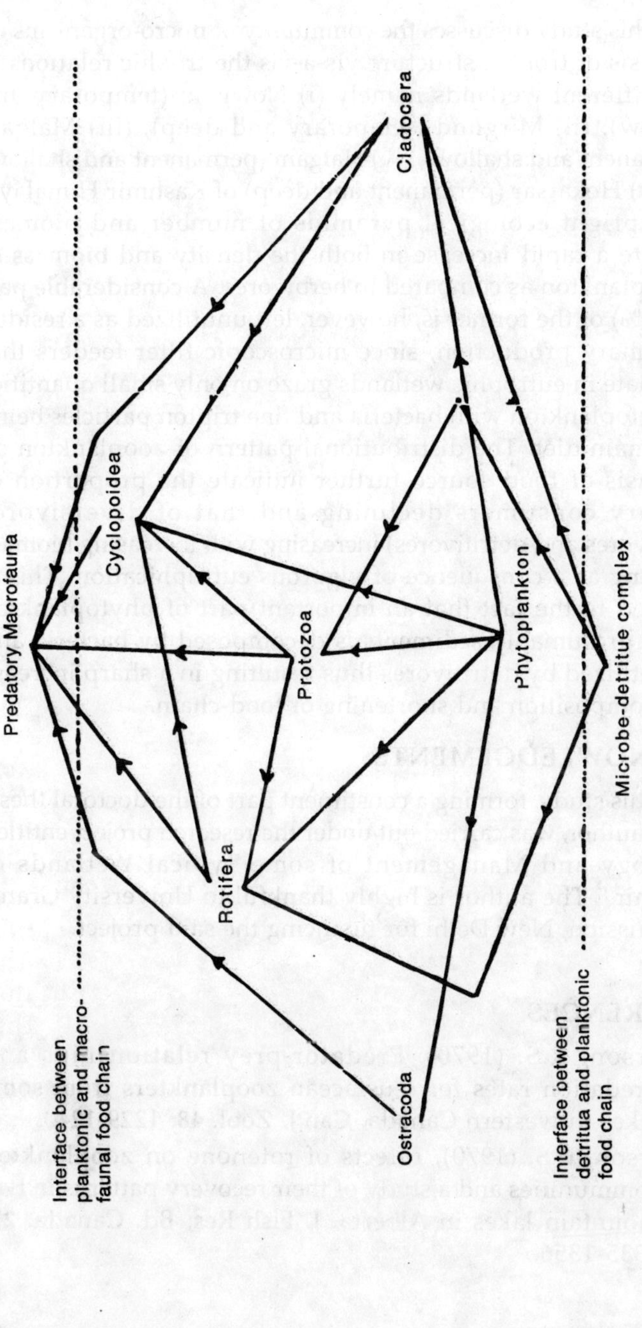

FIG. 7 : A TYPICAL FOOD WEB OF PLANKTON COMMUNITY IN KASHMIR WETLANDS SHOWING THE POSSIBLE FLOW OF MATERIALS (PANDIT, 1980)

CONCLUSION

This study discusses the community of micro-organisms on the basis of trophic structure vis-a-vis the trophic relations in five different wetlands namely (i) Nowgam (temporary and shallow), (ii) Mirgund (temporary and deep), (iii) Malgam (permanent and shallow), (iv) Haigam (permanent and shallow) and (v) Hokarsar (permanent and deep) of Kashmir Himalaya. The upright ecological pyramids of number and biomass indicate a rapid increase in both the density and biomass of phytoplankton as compared to herbivores. A considerable part (70-80%) of the former is, however, left unutilized as a residue of primary production, since microscopic filter feeders that dominate in eutrophic wetlands graze on only small quantities of phytoplankton with bacteria and fine tripton particles being their main diet. The distributional pattern of zooplankton on the basis of food source further indicate the proportion of primary consumers declining and that of diversivores (omnivores and detritivores) increasing with increasing biomass resulting as a consquence of vigorous eutrophication. This is ascribed to the fact that an important part of phytoplankton which presumably sediments is decomposed by bacteria and then utilized by detritivores thus resulting in a sharp increase in decomposition and shortening of food-chain.

ACKNOWLEDGEMENTS

This study, forming a constituent part of the doctoral thesis of the author, was carried out under the research project entitled "Ecology and Management of some Typical Wetlands of Kashmir." The author is highly thankful to University Grants Commission, New Delhi for financing the said project.

REFERENCES

Anderson, R.S. (1970), Predator-prey relationships and predation rates for crustacean zooplankters from some lakes in western Canada. Can.J. Zool. **48**: 1229-1240.

Anderson, R.S. (1970), Effects of rotenone on zooplankton communities and a study of their recovery patterns in two mountain lakes in Alberta. J. Fish Res. Bd. Canada. **27**: 1335-1356.

Anderson, R.S. (1972), Zooplankton composition and change in an alkaline lake. Verh. Internat. Verein. Limnol. **8**: 264-268.

Anderson, R.S. (1974), Crustacean plankton communities of 340 lakes and ponds in and near the National Parks of the Candian Rocky Mountains. J. Fish. Res. Board Can. **31**: 855-869.

Anderson, R.S. and R.B. Green (1975), Zooplankton and phytoplankton studies in the western lakes, Alberta, Canada. Verh. Internat. Verein. Limnol. **19**: 571-579.

Anderson, R.S. and R.B. Green (1976), Limnological and planktonic studies in the Waterton lakes, Alberta. Can. Wildl. Serv. Occasional paper No. 27, 40p.

Brook, J.L. (1969), Eutrophication and changes in the composition of zooplankton. *In*: Eutrophication– Causes, Consequences, Correctives. National Academy of Sciences, Washington, D.C. 236-255.

Brynildson, O.M. and J.J. Kempinger (1973), Production, food and harvest of trout in Nebish lake, Wisconsin. Wis. Dep. Nat. Resour. Tech. Bull. **65**: 1-20.

Carl, C.C. (1940), The distribution of Cladocera and free-living Copepoda in British Columbia. Ecol. Monogr. **10**: 55-110.

Davis, C.C. (1961), The biotic community in the Great lakes with respect to pollution. Proc. Conf. on Water Pollution and the Great Lakes. Depaul Univ., Chicago: 80-87.

Davis, C.C. (1964), Evidence for the eutrophication of Lake Erie from phytoplankton records. Limnol. Oceanogr. **2**: 275-283.

Enright, J.T. (1969), Zooplankton grazing rates estimated under field conditions. Ecology **50**: 1070-1075.

Gliwicz, Z.M. (1969a), The share of algae, bacteria and trypton in the food of the pelagic zooplankton in lakes with various trophic characteristics. Bull. Acad. Poll. Sci. Ser. Sci. Biol. **17**: 159-165.

Gliwicz, Z.M. (1969b), The food source of lake zooplankton. Ecol. Pol. Ser. B. **15**: 205-233.

Gliwicz, Z.M. (1969c), Utilization of primary production by plankton consumers depending upon the length of the food chain. Ecol. Pol. Ser. B. **15**: 63-70.

Gliwicz, Z.M. (1969d), Studies on the feeding of pelagic zooplankton in lakes with varying trophy. Ekol. Pol. Ser. A. **17**: 663-708.

Gliwicz, Z.M. and A. Hillbright-Ilkowska (1972), Efficiency of the utilization of nannoplankton primary production by communities of filter feeding animals measured in-situ. Verh. Internat. Verein. Limnol. **18**: 197-203.

Gulati, R.D. (1975), A study on the role of herbivorous zooplankton community as primary consumers of phytoplankton in Dutch lakes. Verh. Internat. Verein. Limnol. **19**: 1202-1210.

Handoo, J.K. (1978), Ecological and production studies of some typical wetlands of Kashmir. Ph.D. thesis, University of Kashmir, Srinagar.

Haney, J.F., (1971), An in-situ method for the measurement of zooplankton grazing rates. Limnol. Oceanogr. **16**: 971-977.

Haney, J.F. (1973), An in-situ examination of the grazing activities of natural zooplankton communities. Arch. Hydrobiol. **72**: 87-123.

Hillbright-Ilkowska, A. et al. (1971), The variation of some ecological efficiencies and production rates in some plankton community of several lakes of different trophy. *In*: Proceedings of IBP-UNESCO Symp. on "Productivity Problems of Freshwaters" Kazimierz Dolny : 111-127.

Hutchinson, G.E. (1967), A treatise on Limnology. II. Introduction to Lake Biology and Limnoplankton. New York. 1115 p.

Jarnefelt, H. (1952), Plankton als indikator der trophiegruppen der seen. Ann. Acad. Scient. Fennicae. (A iv) 18: 1-29.

Jarnefelt, H. (1956), Zurlimnologie einiger Gewasser Finnlands. XVI. Mit. besonderer Berucksichtigung des planktons. Ann. Zool. Soc. Zool. Bot. Fennicae Vanamo **17**(1): 1-201.

Jarnefelt, H. (1963), Zur Limnologie einiger Gewasser Finnlands XIX. Ann. Zool. Soc. Zool. Bot. Fennicae Vanamo **24**(7): 1-118.

Jumppanen, K. (1976), Effects of waste waters on a lake ecosystem. Ann. Zool. Fennici. **13**: 85-138.

Kaul, S. (1980), Contribution to the analysis of some wetland ecosystems with special reference to biogeo-chemical cycling. Ph.D. thesis, University of Kashmir, Srinagar-190006, India.

Kaul, S. (1982), Community architecture, biomass and production in some typical wetlands of Kashmir. Indian J. Ecol. **9**(2): 320-329.

Kaul, S. and C.L. Trisal (1984), Chemical and physical characteristics of some wetland waters of Kashmir. Acta. hydrochim et hydrobiol. **12**(2): 137-144.

Kaul, V., D.N. Fotedar, A.K. Pandit and C.L. Trisal (1978), A comparative study of plankton populations of some typical fresh waterbodies of Jammu and Kashmir State. *In*: D.N. Sen and R.P. Bansal (Eds.) Environmental Physiology and Ecology of Plants, B. Singh, M. Pal Singh, Dehra Dun, India: 249-269.

Kaul, V. and A.K. Pandit (1979), Is Dal lake dying? Sci. Rep. **16**(11): 734-736.

Kaul. V. and A.K. Pandit (1982), Biotic factors and food chain structure in some typical wetlands of Kashmir. Poll. Res. **1**(1-2): 49-54.

Kaul, V., A.K. Pandit and D.N. Fotedar (1980), Ecology of freshwater snails (gastropod molluscs) in Haigam, a typical wetland of Kashmir Tropical Ecology, **21**(1): 32-46.

Kaul, V. and A.K. Pandit (1980), Management of wetland ecosystems as wildlife habitats in Kashmir. *In*: B. Patel (Ed.) Proceedings of the International Seminar on Management of Environment. Bhabha Atomic Research Centre, Bombay, India : 31-53.

Kaul, V. and A.K. Pandit (1981), Benthic communities as indicators of pollution with reference to wetland ecosystems of Kashmir. *In*: A.R. Zafar *et al.* (Eds.) WHO Workshop on Biological Indicators and Indices of Environmental Pollution. Cent. Bd. Prev. Cont. Poll. Osm. Univ. Hyderabad, India: 33-52.

Kaul, V., C. L. Trisal and J.K. Handoo (1978), Distribution and production of macrophytes in some waterbodies of Kashmir. *In*: J.S. Singh and B. Gopal (Eds.), Glimpses of Ecology (Professor R. Mishra commemoration volume). Intern. Sci. Publ., Jaipur, India: 313-334.

Keeney, D.R. (1973), The nitrogen cycle in sediment-water

systems. J. Environ. Quality, **2**(1): 15-29.

Lohmann, H. (1908), Unter such ungen zur Festellung des Vollatandiger Gehaltes des Meeres an plankton. Wiss. Meeresunters 10.

Macarthur, R. (1955), Fluctuations of animal populations and a measure of community stability. Ecology **36**: 533-536.

Margalef, R. (1961), Communication of structure in planktonic populations. Limnol. Oceanogr. **6**: 124-128.

Moyle, J.B. (1945), Some chemical factors influencing the distribution of aquatic plants in Minneesota. Amer. Midlad. Nat. **34**: 402-420.

Naulapaa, A. (1966), Ereaden suomessa esiintyvien plankter eiden tilavuuksien keskiar voja. Vesiensuo jelutoimiston Tiedonantoja **21**: 1-163.

Nauwerck, A. (1963), Die Beziehungen Zwischen Zooplankton und Phytoplankton in See Erken. Symbolae Bot. Upsaliensis, **17**(5): 1-163.

Ohle, W. (1934), Chemische und Physikalische Untersuchungen Norddeutscher Seen. Arch. Hydrobiol. **26**: 386-464.

Otsuki, A. and R.G. Wetzel (1972), Coprecipitation of phosphate with carbonates in a marl lake. Limnol. Oceanogr. **17**: 762-767.

Pandit. A.K. (1980), Biotic factor and food chain structure in some typical wetlands of Kashmir. Ph.D. thesis, University of Kashmir, Srinagar-6.

Pandit, A.K. (1984), Role of macrophytes in aquatic ecosystems and management of freshwater resources. J. Environ. Manage. **18**: 73-88.

Pandit, A.K. (1988), Threats to Kashmir wetlands and their wildlife resources. Environmental Conservation **15**(3): 266-268.

Pandit A.K. (1991), Conservation of wildlife resources in wetland ecosystems of Kashmir, India. J. Environ. Manage. **33**: 143-154.

Pandit, A.K. and D.N. Fotedar (1982), Restoring damaged wetlands for wildlife. J. Environ. Manage. **14**: 359-368.

Pandit, A.K. and V. Kaul (1981), Trophic structure of some typical wetlands. *In*: B. Gopal, R.E. Turner, R.G. Wetzel and D.F. Whigham (Eds.), Wetlands – Ecology and Management

(Part II). Intern. Sci. Publ. and National Institute of Ecology, India : 55-82.

Patalas, J. and K. Patalas (1966), The crustaceans plankton communities in Polish lakes. Verh. Internat. Verein. Limnol. **16**: 204-215.

Patalas, K. (1968), Crustacean plankton of some Masurian lakes characterized by different fish yield. Pol. Arch. Hydrobiol. **15**: 289-304.

Patalas, K. (1972), Crustacean plankton and the eutrophication of St. Lawrence Great Lakes J. Fish. Res. Board Canada **29**: 1451-1462.

Pattens, B.D. (1962), Species diversity of net phytoplankton in Raritan Bay. J. Marine Sci. **20**: 57-75.

Pejler, B. (1964), Regional–ecological studies of Swedish freshwater zooplankton. Zool. Bidrag. Uppsala. **36**: 407-515.

Rawson, O.S. (1956), Algal indicators of trophic lake typea. Limnol. Oceanogr. **1**: 18-25.

Richards, R.S., C.R. Goldman, T.C. Frants and R. Wickwire (1975), Where have all the Daphnia gone? The decline of a major cladoceran in Lake Tahoe, California-Nevada. Verh. Internat. Verein. Limnol. **19**: 835-842.

Shushkina, E.A. (1966), The estimation of zooplankton production in different lakes. Verh. Internat. Verein. Limnol. **16**: 419-424.

Spodniwwska, I. (1971), The influence of experimental increase of biomass of blue-green algae *Gloeotricha echinulata* (Smith) Richter, on phytoplankton production. Ecol. Pol. **19**: 475-483.

Thienemann, A. (1939), Grundzuge liner allgemeinen Okologie. Arch. Hydrobiol. **35**: 267-285.

Thunmark, S. (1945), Die abwasserfrage der vaxjo seen in hydrobiologischer Belenchtung. Meddel. Lunds Univ. Limnol. Inst. **4**: 1-239.

Tundisi, T.M. and J.G. Tundisi (1976), Plankton studies in a lacustrine environment. Oecologia (Beri.) **25**: 265-270.

Utermohl, K. (1958), Zur vewollkonung der quantitiven phytophankton methodik. Mitt. Int. Ver. Limnol. **9**: 1-38.

Vollenweider, R.A. (1969), A manual on methods for measuring primary production in aquatic environments. IBP Hand Book No. 12. Blackwell Scientific Publications, Oxford: 213 p.

Widuto, J. (1977), Zooplankton in the Kortowskii lake after 18 years of experimental restoration of the lake. Pol. Arch. Hydrobiol. **24**(1): 91-106.

Winberg, G.G. (1972), Etudes sur le bilan biologique energetique et la productive des lacs on Union Sovietique. Verh. Internat. Verein. Limnol. **18**: 39-64.

Wright, J.C. (1958), The limnology of Canyen Ferry Reservoir. I. Phytoplankton–zooplankton relationship in the euphotic zone during September and October, 1956. Limnol. Oceanogr. **3**: 150-159.

17

Modelling Seasonal Fluctuation of Aquatic Fauna with Reference to the Environmental Factors in a High Altitude Lake of Peninsular India (The Ootacamund Lake)

P. Geetha, K. Shanthi, R. Jaganathan & S. Balasubramaniam

INTRODUCTION

In recent years man's constantly growing impact upon the natural ecosystem makes it necessary to plan the restructuring of it on local and later on a regional and possibly a global seek. High altitude mountain lakes of peninsular India are supposed to be particularly subjected to seasonal changes. Because of small drainage areas with limited chemical and biological erosion and extreme climatic conditions, they develop simple and labile ecosystems which react promptly to environmental stress. Ooty lake become eutrophicated in the late 1980s. But the State Government has taken proper measures in eradicating the weeds and the lake is clear (Rao 1994), eventhough the restoration of the lake is still vague. So the present study focuses on the multivariate analysis of impacts on biological factors (Phytoplankton, Zooplankton and micro-organisms) with reference to the environmental factors in the lake.

Canonical correlation is a multivariate statistical technique which is widely used in the study of ecological communities especially water quality studies which reveal the relationship between two groups of variable. (Parameters).

The cluster analysis is another multivariate statistical

technique applied to analyse the correlations between the physico chemical and biological parameters and grouping them into clusters.

MATERIALS AND METHODS

(a) Description of the study area

Ooty lake is located at an altitude of 2500 meter above the MSL. This lake was formed artificially by converging six or seven wild mountain streams during 1823-25 by John Sullivan, the then Collector of Coimbatore. The lake stretches in an irregular 'L' shape with an area of about 4 sq km located at the south western part of the Udagamandalam Municipal Limit in centre of the Nilgiris plateau. The maximum depth is 15 feet and the circumference is 25,000 metres. Ooty has a temperate climate and during summer the mercury level seldom rises above 25°C. The geographical location of the lake provides an ideal ecosystem for the study on the use, distribution and fate of pollutants from the Urban area of Ooty.

(b) Collection of Sample and Analysis

The water sample was collected fortnightly from April to October 1994. The collection was made throughout in a particular spot around 1200 hours to maintain uniformity.

The phytoplanktons and zooplanktons were identified under mircroscope and they were counted separately. Micro organisms were separated according to the method described by Buchanan and Gibbons (1974).

The physico chemical parameters of the water samples collected were analysed using the standard methods, (APHA. 1982).

RESULTS AND DISCUSSION

(a) Introduction of Canonical Correlation

Canonical Correlation was first proposed by Hotelling (1936). In this, we obtain two linear combinations, i.e., one combination of the 'p' predictor variables and one combination of outcome measures $Y_1, Y_2, \ldots\ldots\ldots Y_g$. The linear combinations have vectors a and b. The values of maximum p and r are known as Canonical Correlation R_c between the p and q sets of variables and $X^{(1)}$ and $X^{(2)}$ are the respective Canonical variables. The coefficient of the p and q variables is the Canonical Coefficients. (Takeuchi *et al.*, 1981).

(i) The Model

Analysis of Canonical Correlation often reveals relationship between two groups of variables. The first group of variables (q) have a random vector $X^{(1)}$ and $X^{(2)}$ respectively. The covariance of $X^{(1)}$ and $X^{(2)}$ are :

$$E(X^{(1)}) = \mu^{(1)};\ COV\ (X^{(1)}) = \Sigma\ 11 \qquad \text{....(1)}$$

$$E(X^{(2)}) = \mu^{(2)};\ COV\ (X^{(2)}) = \Sigma\ 22 \qquad \text{....(2)}$$

$$COV\ (X^{(1)}, X^{(2)});\ \Sigma\ 12 = \Sigma\ 21$$

For convenience the random vector $X^{(1)}$ and $X^{(2)}$ are put jointly. By partitioning of the covariance matrix, we can partition p variables.

$$\Sigma = \left[\frac{\alpha^{(1)}}{X^{(2)}}\right] = \left[\begin{array}{c} X_1^{(1)} \\ X_2^{(1)} \\ \vdots \\ X_p^{(1)} \\ \hline X_1^{(2)} \\ X_2^{(2)} \\ \vdots \\ X_q^{(2)} \end{array}\right] \begin{array}{l} \left.\vphantom{\begin{array}{c} X \\ X \\ \vdots \\ X \end{array}}\right\} q \\ \left.\vphantom{\begin{array}{c} X \\ X \\ \vdots \\ X \end{array}}\right\} p-q \end{array} \qquad \text{...(3)}$$

which has a mean vector.

$$\frac{\mu}{(p \times q \times 1)} = E(X) = \frac{E(X^{(1)})}{E(X^{(2)})} \qquad \text{....(4)}$$

and the covariance matrix y (pxq) (p+q) = Σ is

$$\Sigma = E\ (x - \mu)\ \ (X - Y)$$

$$= \frac{\Sigma\ 11 \quad \Sigma\ 12}{\Sigma\ 21 \quad \Sigma\ 22} \qquad \text{....(5)}$$

The pq elements measure the association between the two sets. The linear combination of these variables provide the sum measures of a set of variable as in 1.

(ii) Application of the Model

The physical, chemical and biological parameters of the water body form a complex, mutually cross-correlated entity of information. These complex phenomena cannot be achieved through bivariate statistical analysis and even when achieved the conclusion drawn would not express the real system. Hence, in recent times, the expansion of the use of multivariate statistics in hydro meteorological studies (Crome and Carpenter, 1988; Siver and Hamer, 1989; Varis *et al.*, 1989 and Varis, 1991).

(iii) Results and Discussion

The sample canonical variate coefficient of the first set of canonical variables (Environmental factors) are presented in Table 1. From the results it is observed that the phytoplankton has positive and negative correlations with the first set of canonical variables which may be interpreted as an environmental characteristic index. The pH, conductivity and total N are not closely associated with other environmental factors. The other variables in one way or other shows close association. Similarly the coefficient of second canonical variables (phytoplankton) are also presented in Table 1. There is positive and negative canonical values. The phytoplankton analysis reveals that Anacystis, Chlamydomonus and Clostridium are not associated with one another whereas the other species are closely associated with each other.

The statistical performance of the analysis is shown in Table 2. From the table it is observed that the total N, DO, Volvox and Euglena show poor canonical correlation, the other variables show more or less same correlation. All the species are positively correlated with temperature, pH, turbidity, conductivity, TDS and phosphate.

Varis (1988, 1991) has reported that the application of multivariate analysis in combination with modeled or even descriptive knowledge of the dynamic system provides an effective tool in the elevation of statistical relationship in the process studied. By extracting the correlation structure of

TABLE 1 : RESULTS OF THE COEFFICIENTS FOR CANONICAL VARIABLES
(A) FIRST SET : ENVIRONMENTAL FACTORS

Sl. No.	Temp	pH	Turbidity	Conductivity	Total Solids	Phosphate	Total N	Dissolved Oxygen
1.	0.66681	0.00705	0.19278	– 0.40597	– 1.06746	– 0.08899	0.04559	0.06041
2.	– 0.50556	0.59020	0.30689	– 0.45710	0.24862	0.63003	– 0.53776	– 0.11988
3.	0.20318	– 0.57960	0.31397	– 0.27190	0.13864	– 0.62101	– 0.24755	0.75637
4.	0.42368	– 0.53689	– 0.22926	– 0.41550	0.78911	0.05866	0.40532	– 0.10862
5.	0.09724	– 0.88457	– 0.11910	– 0.45309	– 0.76593	0.11032	– 0.63170	0.36736
6.	0.84126	– 0.18835	– 0.64085	0.37556	0.16808	– 0.00062	– 0.23847	0.26982
7.	0.21497	– 0.10945	0.40022	0.31340	0.80394	– 0.06166	– 0.12347	– 0.82492
8.	0.08301	– 0.06628	– 0.47056	– 0.36286	– 0.30034	– 0.64847	– 0.36690	– 0.39778

(B) SECOND SET : PHYTOPLANKTON

Sl. No.	Anacystis	Oscillatoria	Chlamydomonus	Clostridium	Spirogyra	Ulothrix	Volvox	Euglena
1.	0.57682	0.04353	– 0.06221	– 0.59293	– 0.35324	0.25907	1.00983	0.41326
2.	– 1.02948	0.04768	0.90436	– 0.17695	0.10030	0.12090	– 0.46030	0.59213
3.	– 0.54563	0.41368	– 0.63550	0.68879	0.46821	1.37074	– 0.58484	0.00987
4.	0.34412	– 0.23582	– 0.41960	– 0.23652	0.91047	0.27677	– 0.37034	0.31534
5.	– 1.49450	0.71011	– 0.05141	– 0.16578	0.11571	0.43121	0.16306	– 0.11582
6.	0.96498	– 0.88960	– 0.82166	– 0.68449	– 1.18331	– 0.68494	– 0.27853	– 0.44816
7.	– 0.78813	– 0.50526	0.10055	– 0.04052	0.70524	0.04803	– 0.09711	– 0.83517
8.	– 0.25340	0.29617	– 0.25819	1.36932	0.63270	0.82579	0.46022	0.70770

TABLE 2 : RESULTS OF THE CANONICAL CORRELATION

Sl. No.	Eigen Value	Canonical Correlation	Wilks Lambda	Chisquare	D.F.	Sign Level
1.	1.00000	1.00000	0.00000	143.48	64	0.0000
2.	1.00000	1.00000	0.00000	130.10	49	0.0000
3.	1.00000	1.00000	0.00000	143.48	36	0.0000
4.	0.69660	0.83460	0.06390	12.38	25	0.9833
5.	0.63490	0.79680	0.21070	7.01	16	0.9731
6.	0.36450	0.60370	0.57700	2.47	9	0.9816
7.	0.09140	0.30240	0.90800	0.43	4	0.9796
8.	0.00060	0.02540	0.99940	0.00	1	0.9570

numerous cross concluded variables it allows them to be expressed in a very condensed form.

(b) Introduction of Cluster Analysis

One of the major problems of hydrochemical investigation in hydrogeology is the case with which large quantities of the data are generated. This cluster analysis is only applicable for run off stream waters. However, there are certain constraints in lentic water, especially lakes which have very limited input sources. The major input sources are runoff from adjoining slopes and the natural rain within the water spread area. These features will in one way or other affect the natural water present in the lake.

(i) The Model

The model applied for the study of computer application is agglonurative. The cluster begins with a collection of the number of species in each sampling period which will progressively build groups or clusters of similar sampling times. The subsequent clustering cycles, only pair of entities may be joined to form a new cluster.

The general equation computed for cluster analysis by Lance and Williams (1967) and the linear combinations given as :

D (j, k) (h) = 1 (Dj, h) + 2 D (k, h) + D (j, k) 3 (1)

where j and k are the distance between jth and kth sampling period.

1 and 2 are the parameters.

The 1 and 2 are different in different strategies of cluster analysis which are used for weighing the distance.

(ii) Application of the Model

In order to accommodate the increasing number of physico-chemical parameters in different sampling periods in various seasons, multivariate techniques have been applied for such studies. We have applied cluster analysis on the collected data of phytoplankton in 14 different continuous sampling times. There are 18 phytoplankton species. The relative function of Euclidian distance, which is a resemblance function presented in Table 3 and the pattern of the dendrogram is also presented in Fig. 1.

(iii) Results and Discussion

The clustering cycles are presented in Table 3a. From the table it is assumed that in cluster one sampling periods (3) and (10) are forming a cluster i.e., 1st May and August 15th samples. In the subsequent cycles sampling period (6) is added to above cluster, which falls on June 15th, while using 25.18 as an arbitrary cutoff cluster level, we can observe from the figure, three distinct clusters. The first group forms the cluster of the sampling periods 3, 5, 6, 9, 10, 11 and 12. The second group consists of the sample periods 2 and 13. Whereas group 3 constitutes sampling periods 1 and 4. From the cluster it is concluded that the phytoplankton distribution and abundance are not under seasonal influence.

The above said method is also applied to the microorganisms of the Ooty lake and the Euclidean distance is presented in Table 4a. Based on the Euclidean distance and the cluster level presented in the Table 4b, a dendrogram, is also constructed and the same is presented in Figure 2. From the figure it is assumed that there are three distinct clusters. The first cluster consists of the sampling periods 2,10,3,8,4 and 9 with an arbitrary cluster level of 15.7. The second cluster comprises of 5, 6 and 14 sampling times which has a cut off cluster level of 17.35. The remaining sampling periods i.e., 1,11,7,12 and 13 form a third cluster. From the results presented in Table 4a and 4b and Figure 2, it is observed that the distribution of microorganisms is not under the influence of seasonal effect.

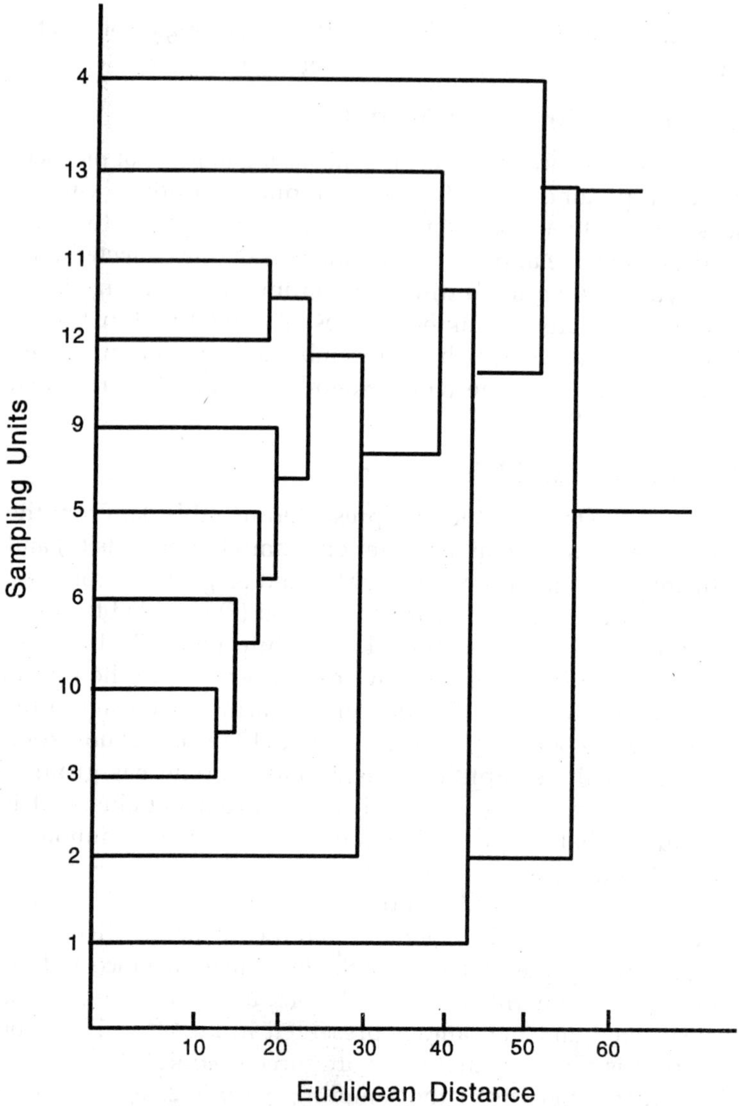

FIG. 1: DENDROGRAM OF THE CLUSTERING OF 14 SAMPLING UNITS (PHYTOPLANKTON) USING EUCLIDEAN DISTANCE

We have also applied the cluster analysis technique to the physico-chemical parameters of the 14 sampling times and relative Euclidean distance are calculated (Table 5a and 5b). The

TABLE 3a : RESULTS OF THE EUCLIDEAN DISTANCE BETWEEN PHYTOPLANKTON COMMUNITY OF OOTY LAKE

0.00	68.28	71.89	70.75	70.97	78.31	67.90	85.71	85.76	61.66	69.62	75.95	76.56	76.95
	0.00	53.55	43.58	41.38	68.32	54.30	40.62	59.55	58.28	43.71	49.62	43.77	43.14
		0.00	65.61	36.11	26.44	54.12	32.74	25.20	41.10	67.41	39.14	32.23	30.25
			0.00	67.81	41.22	29.19	61.98	44.46	28.84	45.83	74.32	45.84	36.00
				0.00	61.25	36.04	18.60	44.47	26.44	23.49	47.98	70.99	31.42
					0.00	44.28	43.08	32.97	36.12	30.53	20.06	52.47	52.52
						0.00	44.06	62.79	68.48	75.93	72.08	65.96	69.84
							0.00	67.07	52.77	43.38	53.12	44.62	41.50
								0.00	59.31	35.54	20.78	41.35	20.62
									0.00	66.85	45.67	29.51	36.48
										0.00	61.58	54.57	50.58
											0.00	63.17	35.11
												0.00	59.32
													0.00

TABLE 3b : RESULTS OF THE CLUSTERING OF THE SAMPLING TIMES (PHYTOPLANKTON)

Cluster	No. Groups	Cluster level	Ref. SU's	SU's in the group
1.	13	18.60	3	10
2.	12	17.49	3	6 10
3.	11	17.60	3	5 6 10
4.	10	22.08	3	5 6 9 10
5.	9	25.18	3	5 6 9 10 12
6.	8	24.99	3	5 6 9 10 11 12
7.	7	28.36	2	3 5 6 9 10 11 12
8.	6	30.94	2	3 5 6 9 10 11 12 13
9.	5	40.07	2	3 5 6 7 9 10 11 12 13
10.	4	40.09	2	3 4 5 6 7 9 10 11 12 13
11.	3	43.36	1	2 3 4 5 6 7 9 10 11 12 13
12.	2	37.38	1	2 3 4 5 6 7 8 9 10 11 12
13.	1	59.03	1	2 3 4 5 6 7 8 9 10 11 12
14.	1	59.03	1	All Su's form one group

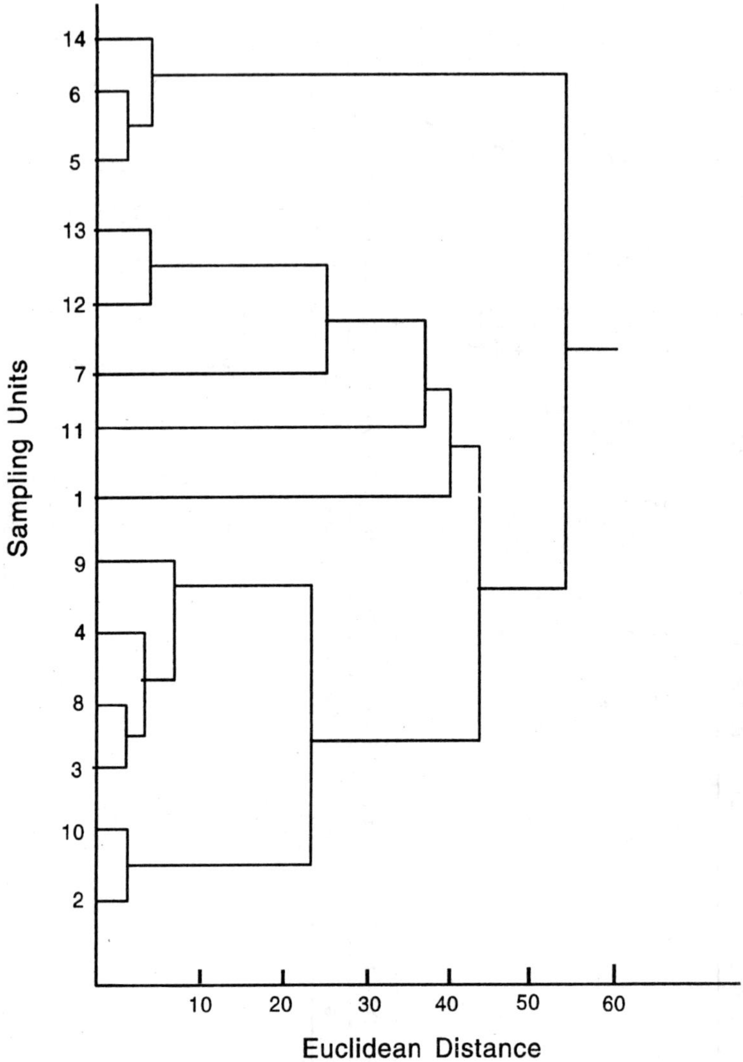

FIG. 2 : DENDROGRAM OF THE CLUSTERING OF 14 (MICRO-ORGANISMS) SAMPLING UNITS IS USING EUCLIDEAN DISTANCE

results of the centroid weighed strategy are summarised in a dendrogram and is presented in Fig. 3. We have used an arbitrary cluster level of 36 to distinguish different clusters. At this level, 3 different cluster groups are formed. The first group

TABLE 4a : RESULTS OF THE EUCLIDEAN DISTANCE BETWEEN MICRO-ORGANISM COMMUNITY OF OOTY LAKE

0.00	145.72	91.11	81.33	67.05	19.52	22.11	101:58	77.99	70.35	95.99	116.32	135.91	133.47
	0.00	34.77	49.91	54.20	69.08	123.29	118.48	34.81	59.56	66.25	45.66	22.93	20.05
		0.00	42.01	49.04	56.97	72.32	126.09	120.95	37.71	59.00	69.74	45.79	22.23
			0.00	43.99	33.78	38.47	53.14	106.39	101.43	18.81	40.06	51.47	32.33
				0.00	62.00	8.94	26.29	36.04	87.06	82.10	22.96	25.69	31.24
					0.00	79.72	28.43	17.97	16.09	61.60	57.46	36.70	20.98
						0.00	75.13	20.20	12.85	17.92	68.45	63.55	26.29
							0.00	55.00	22.74	26.15	37.34	90.54	86.09
								0.00	131.19	76.93	67.30	53.79	13.86
									0.00	135.82	82.02	72.45	57.11
										0.00	83.89	31.58	19.80
											0.00	69.75	23.99
												0.00	65.91
													0.00

TABLE 4b : RESULTS OF THE CLUSTERING OF THE SAMPLING TIMES (MICROORGANISMS)

Cluster	No. Groups	Cluster level	Ref. SU's	SU's in the group
1.	13	8.94	2	10
2.	12	12.85	3	8
3.	11	13.86	5	6
4.	10	15.65	3	4 8
5.	9	12.27	3	4 8 9
6.	8	17.35	5	6 14
7.	7	26.15	3	7 8 9
8.	6	40.20	12	12 13
9.	5	658.75	2	3 4 8 9 10
10.	4	724.50	7	11 12 13
11.	3	1707.88	1	7 11 12 13
12.	2	2250.82	1	2 3 4 7 8 9 10 11 12 13
13.	1	84.39	1	2 3 4 5 6 7 8 9 10 11 12
14.	1	84.39	1	All Su's form one group

TABLE 5a : RESULTS OF THE EUCLIDEAN DISTANCE BETWEEN PHYSICAL AND CHEMICAL PARAMETERS OF OOTY LAKE

0.00	45.05	32.80	42.79	62.41	35.92	31.24	63.23	45.44	64.40	34.03	36.98	33.17	38.85
	0.00	36.53	27.23	58.75	89.98	43.92	44.17	34.39	39.25	33.48	46.21	37.00	31.06
		0.00	40.30	17.13	35.39	65.72	26.48	30.82	49.78	32.05	45.74	37.44	14.03
			0.00	39.86	19.44	32.27	65.18	24.81	28.98	52.88	26.03	50.76	14.76
				0.00	29.67	25.14	31.48	56.55	15.20	11.25	54.80	26.24	56.73
					0.00	44.69	38.69	77.12	10.74	58.56	61.15	17.30	48.96
						0.00	26.38	29.26	46.84	73.14	27.94	28.13	44.40
							0.00	38.19	40.19	79.49	110.20	58.39	58.36
								0.00	28.13	29.82	30.92	53.36	20.50
									0.00	36.60	27.10	27.51	55.57
										0.00	80.35	75.03	36.92
											0.00	55.30	43.92
												0.00	32.22
													0.00

TABLE 5b : RESULTS OF THE CLUSTERING OF THE SAMPLING TIMES

Cluster	No. Groups	Cluster level	Ref. SU's	SU's in the group
1.	13	11.25	6	10
2.	12	14.76	11	12
3.	11	14.59	2	11 12
4.	10	15.04	5	6 10
5.	9	16.42	2	5 6 10 11 12
6.	8	17.30	7	9
7.	7	18.92	2	5 6 8 10 11 12
8.	6	20.97	1	2 5 6 8 10 11 12
9.	5	27.38	1	2 5 6 8 9 10 11 12 13
10.	4	23.43	1	2 5 6 7 8 9 10 11 12 13
11.	3	36.92	3	4
12.	2	32.25	1	2 3 4 5 6 7 8 9 10 11 12
13.	1	41.07	1	2 3 4 5 6 7 8 9 10 11 12
14.	1	41.07	1	All Su's form one group

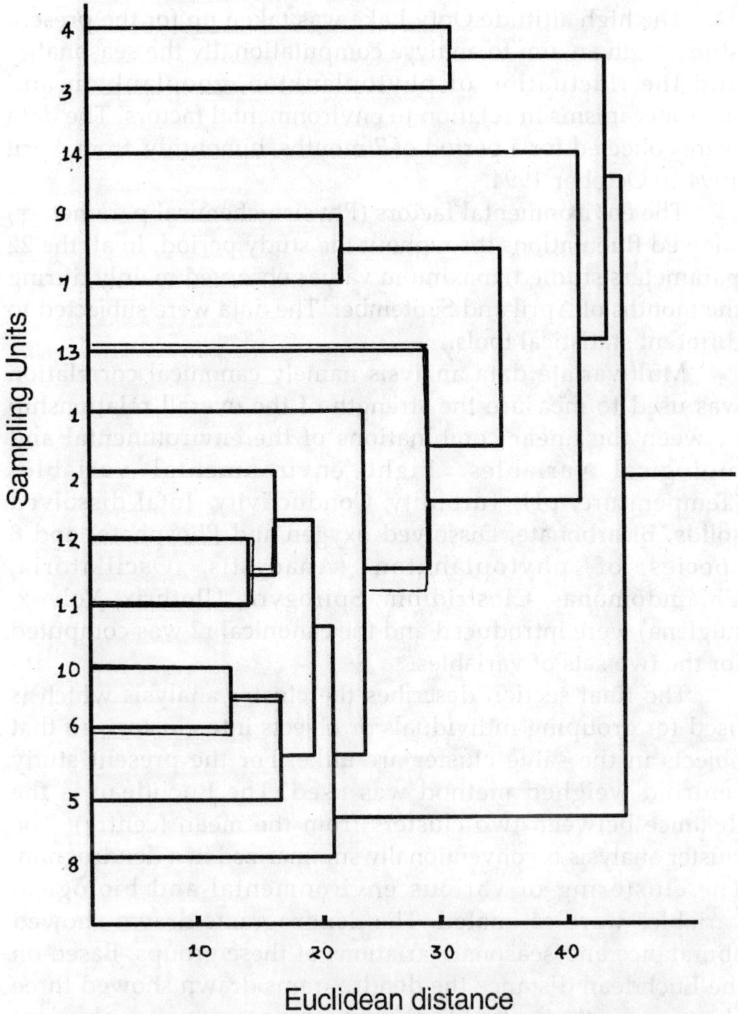

FIG. 3 : DENDROGRAM OF THE CLUSTERING OF 14 PHYSICAL & CHEMICAL SAMPLING UNITS USING EUCLIDEAN DISTANCE

consists of the following chemical groups 8,5,6,10,11,12,2. The second group consists of 1,13,7 and 9 while 14,3 and 4 form the third group.

SUMMARY

The high altitude Ooty Lake was taken up for the present study, with an aim to analyse computationally the seasonality and the fluctuation of phytoplankton, zooplankton and microorganisms in relation to environmental factors. The data were collected for a period of 7 months, bimonthly, from April 1994 to October 1994.

The environmental factors (Physico-chemical parameters) showed fluctuations throughout the study period. In all the 22 parameters studied, maximum values observed mainly during the months of April and September. The data were subjected to different statistical tools.

Multivariate data analysis namely canonical correlation was used to measure the strength of the overall relationship between the linear combinations of the environmental and biological variables. Eight environmental variables (Temperature, pH, Turbidity, Conductivity, Total dissolved solids, Bicarbonate, Dissolved oxygen and Phosphate) and 8 species of phytoplankton (Anacystis, Oscillatoria, Chlamdomonas, Clostridium, Spirogyra, Ulothrix, Volvox, Euglena) were introduced and the canonical r2 was computed for the two sets of variables.

The final section describes the cluster analysis which is used for grouping individuals or objects into clusters, so that objects in the same cluster are alike. For the present study centroid weighed method was used. The Euclidean is the distance between two clusters from the mean (centre). The cluster analysis is conventionally summarized in a dendrogram. The clustering of various environmental and biological variables were estimated. The dendrograms drawn showed abundance and seasonal variations of these groups. Based on the Euclidean distance the dendrograms drawn showed three clusters in phytoplankton, micro-organisms and physico-chemical parameters.

CONCLUSION

In general water quality studies now suffer a lot due to the difference between theoretical and applied research. It is well known that Ooty lake is mainly used for recreational purposes and the State Government is incorporating so many ventures to attract tourist from all over India and abroad. The political and socio-economic value of the lake is very high. The present study confirms that the lake is much polluted, the source being the dumping of sewage. Considering the aesthetic quality and recreational value, the lake needs proper management in the eradication of harmful micro-organisms. Not only local people but also the government should take proper action in maintaining the quality of the lake in order to save the tourists from various diseases. This can be achieved only by proper regulation of sewage discharge into the lake.

REFERENCES

APHA, (1982), Standard methods for the examination of water and waste water. American Public Health Association, Washington, 769 p.

Buchanan, R.E., and Gibbons, N.E. (1974), Bergeys Manual of Determinative Bacteriology 8th Edition Williams and Wilkens, Baltimore.

Crome, F.H.J and Carpenter, S.M. (1988), Plankton community Cycling and recovery after drought - dynamics in a basin on a flood plain. Hydrobiologia. **164** : 193-211.

Hotelling, H., (1936) : Relation between two sets of variates. Bimetrika, **28** : 321-377.

Krishnamoorthy, K.P., and Sarkar, R. (1979), Macro invertebrates as indicators of water quality proc. symp. Environ., Biol. 133-138.

Lance, G.N. and Williams, W.T. (1967), A general theory for classificatory sorting strategies. 1. Hierarchical systems. Computer Journal. **9** : 373-380.

Rao, V.N.R., Mohan, R., Hariprasad, V. and Ramasubramanian R. (1994), Sewage pollution in high altitude Ooty lake, Udagamandalam cases and concern. Poll. Res. **13**(2) ; 133-150.

Siver, P.A. and Hamer (1989), Multivariate statistical analysis of the factor controlling the distribution of scaled Chrysophytes. Limnol. Oceanogr **34** : 368-381.

Takeuchi, K., Yanai. H. and Mukherjee, B.N. (1981), Canonical correlation analysis and discriminant analysis. *In*: The foundations of multivariate analysis. Wiley Eastern Ltd. 225-280.

Varis, O. (1988), Temporal sensitivity of Aphanizomenon flosaquae dominance – A whole lake simulation study with input perturbations. Ecol. Modelling **43**: 137-153.

Varis, O., Sirvio, H. and Kettunen. J. (1989), Multivariate analysis of lake phytoplankton and environmental factors Arch. Hydrobiol. **117**: 163-175.

Varis, (1991), Association between lake phytoplankton community and growth factors – a canonical correlation analysis. Hydrobiologia, **210**: 209-216.

18

Biodiscs – The Biological Wheels for Treatment of Polluted/Waste Waters : An Overview

A.K. Verma

Water pollution and the waste disposal without polluting the environment is one of the most difficult task for the scientists world over. The problem is complicated by the broad physico-chemical spectrum of wastes from municipal, domestic or Industrial operations. Despite the fact that modern technology has provided effective methods for waste water/sewage treatment, many municipalities and communities do not employ adequate treatment procedures; but, even worse, some dump the waste-waters into the water-ways without performing any treatment. Thus treatment of polluted/waste water is necessary before its disposal producing, harmful effects.

Treatment and disposal of waste-water and sewage from individual or other unit structures is accomplished by various methods most common being the trickling filters, activated sludge and stabilization ponds etc. The treatment procedures involves the *Primary treatment to remove coarse* and settleable solids, the *Secondary treatment* to remove dissolved and colloidal organic solids by employing the biological processes like *adsorption, coagulation* and *flocculation* of waste reducing BOD (Biochemical Oxygen Demand) and the *Final treatment* to disinfect and dispose of solid wastes and liquid effluent. The resulting degradation and concentration of the waste water constituents produce a highly purified effluent.

Bio-discs commonly known as Rotating Biological Contactors (RBCs) or Biological wheels is a new and relatively recent waste water treatment process whose first commercial application has been reported in 1960s.

HISTORY OF BIO-DISCS

Although Allen (1929) has reported about the "Biological Wheels" and Doman (1929) on the "Rotating Discs" for waste-water treatment in the Journal Water Pollution Control Federation. However, development of modern Biodiscs or RBCs began in Germany with the work of Hartman and Popel who named their units as "Immersion Drip Filters" in 1950s. Subsequently, polystyrene discs were also manufactured by Germans and the first commercial RBC was installed in 1966. Americans started work on RBCs in mid-1980s using the trade named *Bio-discs*. Research on RBCs in Japan began in 1960 and their installation began operational in 1966 and the first public sewage treatment plant was opened in 1978 after seeking government approval.

BIODISCS-STRUCTURE AND PROCESS DESCRIPTION

Biodiscs (RBCs) consists of a number of thick, vertical, circular discs arranged on a rotating horizontal shaft (Figs. 1-A-B). These discs are made up of metals or polystyrene, polyvinyl chlorides (PVC) or fibre-glass, fitted on 20-25 feet long shaft. The discs may be flat, corrugated, honey combed or otherwise formed to increase the surface area.

During operations, the discs are 40% submerged in a tank of waste-water, so that the rotation alternately exposes the surface to the atmosphere above the tank, then resubmerging them in the waste-water. Rotation is normally done by the speed motor at a fixed speed of generally 1.6 rpm. Rotation serves to:

(1) bring the biofilm in immediate contact with waste-water,
(2) expose the bio-film to the air to provide it with oxygen;
(3) aerate the tank contents by increasing the surface turbulence;
(4) keep the tank contents mixed, so that the settleable waste-water solids and "sloughings" will not accumulate in the tank;
(5) limit/check the bio-film thickness by producing shear-forces as the film passes through the liquid in the tank; and

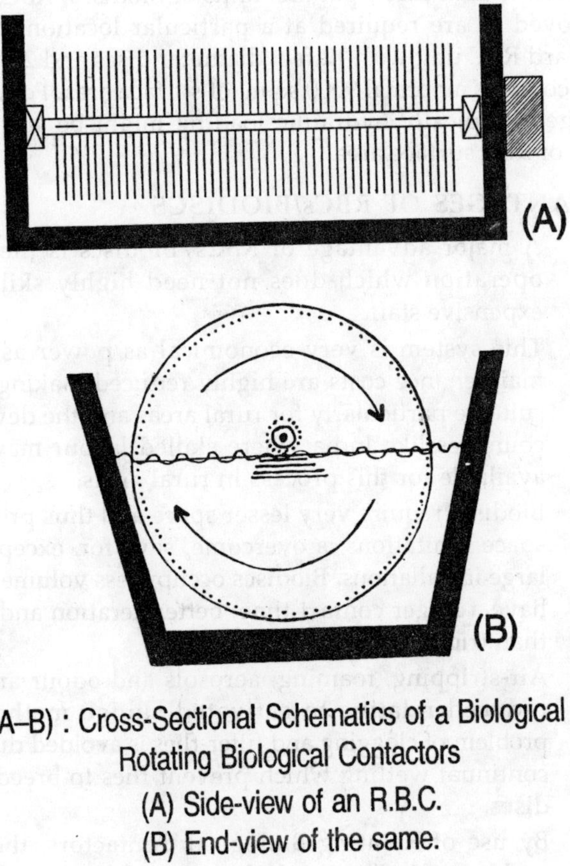

Fig. 1(A–B) : Cross-Sectional Schematics of a Biological Wheel/ Rotating Biological Contactors
(A) Side-view of an R.B.C.
(B) End-view of the same.

(6) increase the amount of the surface area exposed to the tank contents.

Rotations lead to the development of 2-3 mm thick bio-films of microbiol community on to the discs which subsequently derive the nutrition and energy source from the waste water constituents leading to the purification.

During this process, the fragments of bio-film separate periodically from the disc sur.'aces and are carried out of the tank by the waste-water flow to be removed in a secondary settling tank.

The waste-water typically flow through the RBC tank either parallel or perpendicular to the shaft to distribute the load more uniformly over all the discs on the shaft. Depending upon the expected load as many parallel units of Biodiscs/RBCs may be employed as are required at a particular location. A single, standard RBC unit with 12 feet diameter discs and 25 feet long shaft contains a normal 9000 sq.m. of surface area. For domestic sewage, a typically hydraulic loading might be 40 litres/d/sq.m. of RBC surface area.

ADVANTAGES OF RBCs/BIODISCS

I. A major advantage of RBCs/Biodiscs is the ease of operation which does not need highly skilled and expensive staff.

II. This system is very economical as power as well as maintenance costs are highly reduced making it most suitable particularly for rural areas and the developing countries like India where skilled labour may not be available for this process in rural areas.

III. Biodiscs require very lesser space and thus problem of space limitations is overcome, save for, exceptionally large installations. Biodiscs occupy less volume and yet have a larger contact time, better aeration and mixing than trickling filter process.

VI. Air-stripping, foaming, aerosols and odour are much reduced relative to activated sludge method. The problem of clogging and filter-flies is avoided due to the continual wetting which prevent flies to breed on the discs.

V. By use of Rotating Biological Contactors, the major problem of bulking and effluent recycle as encountered with activated sludge and trickling filters respectively is avoided.

VI. In some cases, intermediate settling tanks may not be needed with RBCs particularly for phosphorus removal after RBC and nitrification. Moreover, at higher temperatures, oxygenation of waste-water is easier with biodiscs than the activated sludge.

DISADVANTAGES OF BIODISCS

Corrosion of steel media supports, lack of operational control, troubles with bearings and shaft-failure are some frustrating problems related to the manufacturing and operating experience. Also the polyethylene discs may not be usable at waste-water temperatures over 100°F. Some additional operational problems including uneven loading, solid deposition on tanks, oil leakage from mechanical drives and mineral deposition on the discs are also disadvantages of the biological wheels which are, however, easily overcome with time and experience.

PRESENT STATE OF ART

There were over 5,000 RBC installations world-wide at the beginning of 1980 mostly in the European countries like West Germany, France, Switzerland, Austria, Great Britain, Norway and Italy. Even Yugoslavia since 1974 has many small package plants serving 10-500 people. In Asia, Japan is the leading country having highest number of RBC than reported elsewhere (1,323 installed and 300 under construction till mid-1981). As of 1980, India had no full scale RBC installation, but pilot studies were underway in contrast to other Asian pockets namely Hongkong, South Korea, Taiwan, Dubai and New Zealand having small RBC units. USA have 321 operational installations and numerous planned or under construction units, while Canada has 42 small installations.

CONCLUSION

From what has been said above, it is amply clear that the advantages of RBCs/Biodiscs to Trickling filters, lagoons and activated sludge are numerous. However, with this new technology, certain unanticipated problems may develop which may take some time to get overcome resulting in the achievement of the desired performance by using these *Biological Wheels* in the treatment of *Waste Waters.*

Since domestic and industrial waste waters are the major source/cause of pollution of water bodies in our country, therefore, there is an urgent need to look into this matter very seriously. Therefore, it is suggested that industrial as well as the domestic sewage should be pretreated at the source itself, before the same are discharged into the municipal sewers, for which *Biodiscs* may play a significant role, to some extent, in this regard.

REFERENCES

Allen, K. (1929). The biological wheel. Sewage Works Journal (Journal Water Pollution Reservoir) 1: 560.

Doman, S.W. (1929). Results of operation of experimental contact filter with partially submerged rotating plates. Sewage Works J. 1 : 55.

Hartmann, H. (1950). Untersnehung uber die Biologische Reinigung Von Abwaser Hilfe Von Tanchtropfkorperanlagen (Investigations of the Biological Clarification of Waste Water using Immersion Drip Filters) 249 p.

Strom Peter, F. and Chung, J.C. (1987). The Rotating Biological Contactor for Waste Water treatment. *In*: Advances in Biotechnological Processes 5, Pages 193-225. Alan R. Liss Inc. New York (Ed. Ali Mizradh).

19
Potential Utilization of Aquatic Weeds for Treating Industrial Effluents

A.G. Murugesan & N. Sukumaran

INTRODUCTION

Industrial wastes and discharges have been recognised as one of the major sources of toxic chemicals polluting the environment. Water pollution is one of the most serious problems faced by man today. Since water is the universal enabling chemical which is capable of dissolving or carrying in suspension a variety of toxic materials, several rivers and other aquatic systems are utilized for the disposal of industrial effluents and sewage. Due to rapid industrialization and urbanization, most of our Nation's land and water are deteriorating fastly. Lack of wastewater treatment facilities or their inefficiency is the single greatest cause of pollution. So, there is an urgent need of establishing cost-effective treatment system.

NEED FOR ALTERNATIVE WASTEWATER TREATMENT TECHNOLOGIES

In developed countries, greater capital and technological resources help in building sound collection, treatment and disposal facilities for most of the pollutants whereas the situation is pathetic in developing countries. Only 37 per cent of the sewage gets any treatment in big cities whereas it is only 16 percent in the case of small cities. Out of the 5000 large and medium industries, only 460 had effluent treatment plants and the remaining simply dump their wastes into the environment. It is difficult to separate industrial growth from environmental pollution but it can be minimised through cost effective approaches of pollution abatement. To reach the full objectives of zero pollution, adoption of alternative technologies which suit the situation of low capital availability, minimum manpower and limited energy consumption are necessary. The conventional engineering and mechanical methods of effluent treatment are uneconomical for industries of low turn over rates. In the absence of regular maintenance, the process are delicate and are prone to failure. The use of hydrobotanic treatment for pollution abatement is a good approach and its perspectives are discussed in this paper.

AQUATIC MACROPHYTES

Several aquatic macrophytes such as the water hyacinth, *Eichhornia*, water lettuce, *Pistia*, water fern, *Salvinia* and duck weeds, *Lemna*, *Spirodela* and *Wolffia* are found scavenge inorganic and some organic compounds from wastewaters (Boyd, 1969). *Eichhornia* is a nuisance aquatic weed which was accidentally introduced in West Bengal by queen Victoria during 1889 as an ornamental plant and it spreaded all over India. It is an intransigent weed colonizing vast structures of water bodies with explosive growth rate. Eight plants can multiply to 6 lakhs in eight months. It is estimated that the direct losses due to water hyacinth in developing countries are more than 200 million dollars every year. Similarly, the menace of other weeds like *Pistia*, *Salvinia* etc. also reaching alarming dimensions in many parts of our nation because, they affect water resources adversely by blocking pumps and canals in irrigation projects and interfere with hydroelectricity production. The most important problem is the wastage of water in evapotranspiration in addition to fostering water borne diseases.

MACROPHYTES FOR POLLUTION CONTROL

The very ecological devastating properties of these aquatic weeds make them an ideal candidate for the treatment and nutrient absorption from wastewaters. The treatment by aquatic macrophytes does not involve any elaborated mechanism or setting up a complex unit. The floating nature, their easy availability, fast growth rate, simple growth requirements, their appetite for nutrients and a variety of pollutants, easy harvesting and subsequent economic potential of the harvested plants as feed, fertilizer and biogas make them ideal candidates for wastewater treatment.

POTENTIALS OF EICHHORNIA CRASSIPES IN REMOVING POLLUTANTS

The fantastic ability of absorbing nutrients and rapid growth of water hyacinth, *Eichhornia,* its low economic maintenance and many other profits has raised its value for pollution abatement. Though a large number of laboratory and field trials have been made on treatment of wastewater utilizing water hyacinth, the actual use in the industrial effluent or sewage treatment facilities have been relatively very few. The United States National Aeronautics and Space Administration (NASA) in Missisippi has been using water hyacinth for upgrading wastewater treatment lagoons and to treat chemical wastewaters. The use of this plant in the tertiary treatment system worked well. Regulation of harvesting schedules are needed for maximum absorption of pollutants. Because, the dense mat formation makes the water devoid of life in the water. Young plants are more efficient than older ones. The absorbed heavy metals may be released back to the medium when the plants die off due to over accumulation of toxicants.

Several studies show that water hyacinth reduces suspended particles, dissolved impurities, nitrogen, phosphorous and other nutrients, BOD, COD, turbidity, heavy metals like copper, cadmium, nickel, silver, chromium, iron, zinc, magnesium, manganese, cobalt, lead, mercury, arsenic, several pesticides and other organic pollutants in addition to algae, faecal coliforms and other pathogenic bacteria. During summer, water hyacinth based systems work efficiently due to rich growth rate.

Eichhornia has effectively reduced 97% of BOD, 88% of COD, 75% total solids, 99% of nitrate, 80% of total phosphorous, and considerable quantity of calcium, magnesium, potassium and sodium from sewage in a fortnight (Trivedy and Gudekar, 1985). Studies conducted by Wolverton and Mc Donald, (1975) showed that water hyacinth accumulated cadmium, lead, mercury, nickel, silver and chromium in its body. The roots accumulated more heavy metals. One gram of dry plant material is capable of removing 0.6 mg cobalt, 0.5 mg strontium, 0.4 mg silver, 0.7 mg cadmium, 0.5 mg nickel, 0.2 mg lead and 0.2 mg mercury in 24 hours. So, one hectare of water hyacinth can efficiently remove 263 g silver, 340 g cobalt, 326 g strontium, 300g cadmium or nickel, 105 g lead and 90 g mercury in one day. Increase in the concentration of cadmium in the test solution enhanced the uptake by the roots and tops of *Eichhornia.* When 1 ppm of cadmium solution was treated, the root absorbed 805 mg cadmium per one gram dry weight of the plant in a day whereas it was 2650 mg after 16 days. But in 8 ppm cadmium solution, the roots absorbed 4,444 and 10,600 mg in the first and 16th days respectively (Muramoto and Oki, 1983). When lead and mercury were mixed along with cadmium, the uptake of lead was declined in all the doses and durations. Experiments of Hardey and O' Keeffe (1985) reveal that water hyacinth with greater root mass take up more metal ions as a function of time. The roots possess sites which bind free cadmium. In an experiment, 100 mg dry weight of the roots removed 360 mg cadmium in 24 h whereas 300 mg root mass has removed 400 mg cadmium.

The efficiency of *Eichhornia* in reducing pollution load from industrial effluents are recorded by some investigators. It has reduced 94% of BOD from sugar refinery effluent in one week. When dairy effluents were treated with water hyacinth, 84% BOD, 88% COD, 50% each of total solids, nitrogen and phosphorous were reduced in 14 days. Eventhough they reduced BOD (40%), COD (35%), suspended solids (85%), and ammonical nitrogen (54%) from palm oil mill effluent in 5 days, the plants withered in a few days because of high ammonical nitrogen (Table 1). But when these effluents are treated after partial digestion all the pollutants were removed efficiently. John (1984) reported high reduction of BOD (94%), COD (89%), suspended solids (89%), and ammonical nitrogen (70%) by water hyacinth from natural rubber effluent in 15 days. Tannery effluent required dilution for moderate removal of pollutants as reported by Prasad *et al.* (1983). But it removed cent percent chromium from tannery effluent in 16 days. Similarly, water hyacinth reduced a good quantity of pollutants from paper mill and textile effluents. Heavy metals like copper, nickel and zinc were also reduced from electroplating industrial effluents in 3 days. Murugesan *et al.* (1996b) reported cent percent removal of

TABLE 1 : REMOVAL OF POLLUTANTS FROM EFFLUENTS BY VARIOUS AQUATIC MACROPHYTES

Name of the Macrophyte	Type of Effluent	Retention Time (days)	Parameter	Reduction (%)
Eichhornia	Dairy	7	BOD	87
	Dairy	7	COD	63
	Tannery	7	BOD	35
	Tannery	7	COD	70
	Sugar	1	BOD	24
	Sugar	7	BOD	94
	Distillery	30	Calcium	40
	Palmoil	5	BOD	40
	Palmoil	5	SS	96
	Pulp & paper	15	COD	70
	Pulp & paper	15	SS	80
	Textile	30	Calcium	32
	Textile	30	Magnesium	31
	Mineral	7	Chloride	62
	"	7	Titanium	100
	"	7	Chromium	77
	"	7	Aluminium	87
	"	7	Zirconium	63
	"	7	TS	94
	"	7	Sulphate	60
	Rubber	7	BOD	89
	"	7	COD	79
	"	7	Chloride	84
	"	7	Sulphate	61
	Sewage	7	Nitrogen	75-92
	"	14	COD	60-88
	"	14	TS	75
	"	10	Ammonia	81
	"	14	Nitrate	99
	"	14	Total phosphorous	87
	Pesticide	4	Sodium penta-chlorophenate	82
Pistia	Heavy metal solution	0	Chromium	100
Spirodela	"	2	Chromium	76
	"	2	Nickel	78
Salvinia	"	2	Chromium	92
	"	2	Nickel	90
	"	1	Copper	90
	"	1	Mercury	76

titanium, 88% of aluminium 78% of chromium, 63% of zirconium and 95% of silica, in one week, from mineral factory effluent by water hyacinth.

Since phenol and phenolic derivatives are common organic pollutants found in domestic and industrial wastewaters, they were studied for absorption by *Eichhornia* and found that 2.75 g dry weight of the plant absorbed 100 mg of phenol in 72 hrs. which is equivalent to 160 kg phenol removal by one hectare of water hyacinth (Wolverton and Mc kown, 1976). Presence of copper and zinc along with phenol did not affect the processes and all the three were removed well. According to the reports of Murugesan *et al.* (1996c) water hyacinth reduced 78% sulphate, 96% chloride, 80% BOD, 86% COD, 79% total hardness 88% total dissolved solids and cent per cent total suspended solids in 7 days from rubber factory effluent. But the per cent reduction was less in the case of integrated liquid hydrogen plant effluent in which it has reduced 44% sulphate, 57% chloride, 63% COD, and 58% total dissolved solids in one week. Water hyacinth has reduced the pesticide, sodium pentachlorophenate (64%) and radionuclides Cs and Sr also. Polprasert *et al.* (1992) observed 54-75% removal of BOD, 74-93% of COD, 88% of suspended solids and 50% organic nitrogen from pig wastewater in 10 days. Removal of coliforms, *E. coli* and Streptococcal populations was also recorded.

Experiments were conducted with a hydraulic retention time from 2 to 15 days and satisfactory results obtained in 10 days. It has been found that there was no benefit in diluting the effluent for macrophytes based treatment. The depth of effluent column found efficient was 30 c.m. The pH of 5-7.5 was found favourable because at lower and higher pH, the plants died in a few weeks. It is estimated that 5 ha of water hyacinth is enough to recycle the wastes produced from 10,000 persons. Also it recycles air environment also by higher gas exchange (CO_2-O_2) resulting in an atmospheric revitalization (Gian Gupta, 1982).

OTHER MACROPHYTES FOR POLLUTION CONTROL

Macrophytes like water fern, *Salvinia*, water lettuce, *Pistia*, duck weeds, *Lemna*, *Spirodela*, *Wolffia* and the water peanut, *Alternantherea* were found reducing pollutants from wastewaters. *Salvinia molesta* removed chromium (92%), nickel

(90%) from 1 to 8 ppm solutions in two days. Similarly, *Spirodela* removed more than 75% of these metals. The removal was fast for the first two days of exposure after which it continuously declined due to fast attainment of saturation state in the plants. A slight decline in absorption was noticed when two heavy metals were mixed. *Salvinia* was found more effective than *Spirodela* for heavy metal absorption and such plants can be used to treat wastewater containing low concentration of metals (Srivastav *et al.*, 1994). Table 2 reveals that *Trapa bispinosa* has reduced 74% sulphate, 71% BOD, 75% COD and 80% total dissolved solids, while *Pistia stratiotes* reduced 89% sulphate, 69% chloride, 67% BOD, 72% COD, 90% total hardness and 89% of total dissolved solids. *Salvinia molesta* has proved its efficiency by removing 83% sulphate, 87% chloride, 92% BOD, 78% COD and 82% total dissolved solids from rubber factory effluent in 7 days. *Trapa* has removed 75% sulphate, 72% chloride, 68% COD and 77% total dissolved solids while *Salvinia* reduced 56% sulphate, 50% chloride, 72% COD, 76% total suspended solids and 74% total dissolved solids from integrated liquid hydrogen plant effluent in 7 days (Murugesan *et al.*, 1996c). *Salvinia natans* has absorbed 90 per cent of copper in one day. When the pollutants were at low concentrations, the accumulation occurred only in the roots while, at higher concentration, the shoots also involved in absorption. *Salvinia* has accumulated 90% of mercury in one day from a 5 ppm solution but an increase in mercury concentration beyond 5 ppm make the plant wilted after 3 days.

The duck weeds accumulated heavy metals like lead, chromium, nickel, copper, cadmium, zinc etc. and these metals are precipitated during transfer to clean water. (Timofeeva and Stom, 1988). The shoot system of *Pistia stratiotes* accumulated more potassium, calcium, and magnesium whereas the roots uptake cadmium, chromium, cobalt, iron, lead, mercury, nickel, sodium, manganese and zinc (Sridhar, 1988). The design of macrophytes based treatment systems as three functional zones such as purifying zone, effluent upgrading zone and multiutilization zone gave good results. Macrophytes were planted in several subzones in series in some ponds and these multimacrophyte zone system revealed greater purification efficiencies than those of oxidation ponds. Algal-bacterial zones may also be combined with this.

TABLE 2 : EFFICIENCY OF VARIOUS AQUATIC WEEDS IN REMOVING POLLUTANTS (%) FROM INDUSTRIAL EFFLUENTS (MURUGESAN *et al.*, 1996)

	Eichhornia	*Trapa*	*Pistia*	*Salvinia*
Rubber Factory Effluent				
Sulphate	78	74	89	83
Chloride	96	39	69	87
BOD	80	71	67	92
COD	86	75	72	78
Total Hardness	79	70	90	67
TSS	100	60	78	73
TDS	88	80	89	82
Integrated Liquid Hydrogen Plant Effluent				
Sulphate	44	75	–	50
Chloride	44	75	–	56
COD	63	68	–	72
Total Hardness	57	62	–	75
TSS	49	55	–	76
TDS	58	77	–	74

MECHANISM OF POLLUTANT REMOVAL

The mechanism of purification in macrophyte based treatment systems are very interesting (Fig.1). The presence of macrophytes alters the physico-chemical environment of the water. The dense cover of the plants depletes dissolved oxygen of the underlying water and create anaerobic conditions which favour denitrification process and maximizing nitrate removal. The metal uptake is brought about by rapid transpiration in the plant and the vacuum created effects on upward migration of the liquid medium which is promoted by the energy of incident solar radiation. Stirring of the medium enhanced the pollutant uptake. The macrophytes function as fixed film reactors. Some plants transport oxygen from the foliage into the root and an oxidised environment is created in the water due to the diffusion of oxygen not required for root respiration and this is utilized for bacterial oxidation of nutrients. Biodegradable organic compounds are removed by symbiotic activities of a large number of aerobic bacteria grown on the surface of the roots of the plants. Also, excretion of some organic substances

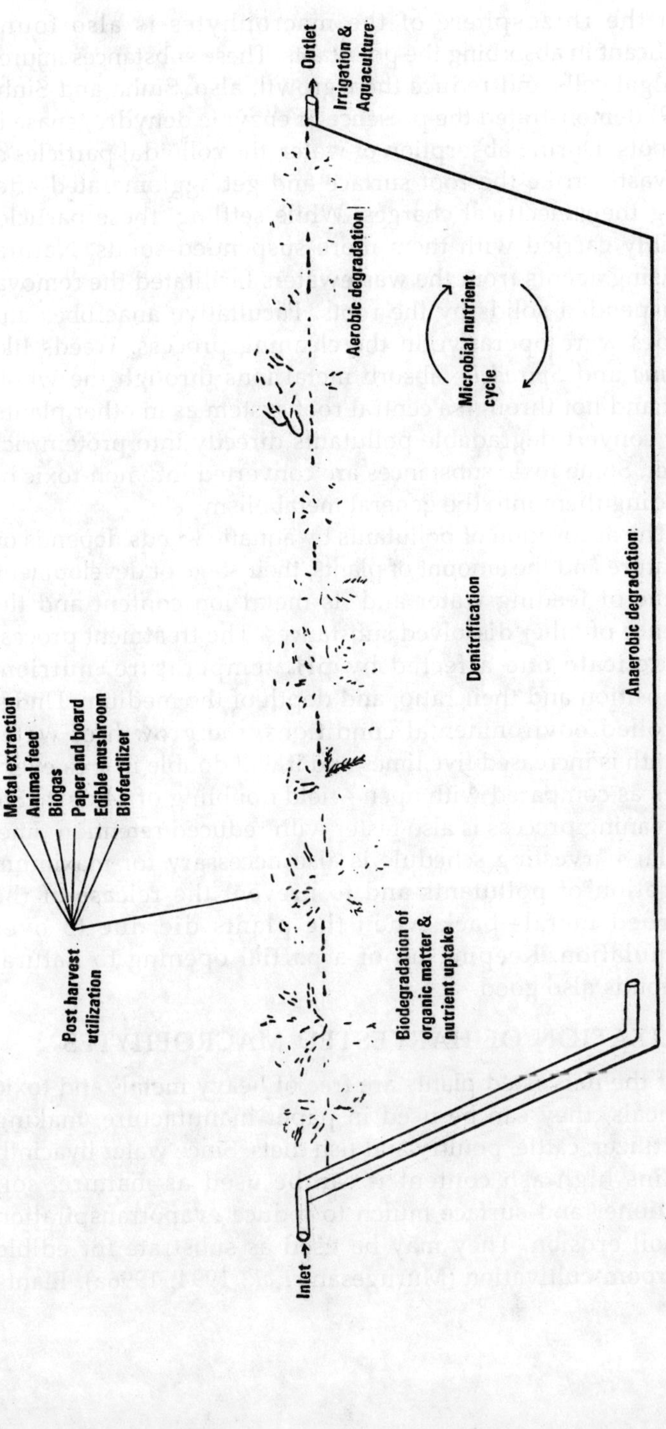

FIG. 1 : PROCESS OF POLLUTANT REMOVAL IN AN AQUATIC MACROPHYTE BASED EFFLUENT TREATMENT SYSTEM

from the rhizosphere of the macrophytes is also found significant in absorbing the pollutants. These substances injured the algal cells and reduce their growth also. Sinha and Sinha (1969) demonstrated the presence of enzyme dehydrogenase in the roots. During absorption of water, the colloidal particles of the waste strike the root surface and get agglomerated after losing their electrical charges. While settling, these particles possibly carried with them more suspended solids. Natural peptising agents from the wastewaters facilitated the removal of suspended solids by the roots. Facultative anaerobes and aerobes were operative in the cleaning process. Weeds like *Salvinia* and *Spirodela,* absorb metal ions through the whole plant and not through a central root system as in other plants. They convert degradable pollutants directly into protein rich fodder. Some toxic substances are converted into non-toxic by including them into the general metabolism.

The absorption of pollutants by aquatic weeds depends on the nature and the amount of plants, their stage of development, volume of feeding water and its metal ion content and the presence of other dissolved substances. The treatment process is a delicate one affected by pH, temperature, nutrient composition and their ratio, and depth of the medium. Under controlled environmental conditions, the growth of water hyacinth is increased five times and it will double in area every 3 days as compared with open - field doubling of 14 days. So, the cleaning process is also faster with reduced retention time. Regular harvesting schedule is also necessary for maximum absorption of pollutants and to prevent the release of the absorbed metals back when the plants die due to over accumulation. Keeping up of a partial opening to natural aeration is also good.

UTILIZATION OF HARVESTED MACROPHYTES

If the harvested plants are free of heavy metals and toxic chemicals, they can be used in paper manufacture, making biofertilizer, cattle, poultry and fish diets. Since water hyacinth contains high ash content it can be used as manure, soil conditioner and surface mulch to reduce evapotranspiration and soil erosion. They may be used as substrate for edible mushroom cultivation (Murugesan *et al.,* 1994, 1996a). Plants containing heavy metals can be used or biogas generation and the slurry can be stored in specially designed pits and after accumulation of metals, they can be extracted. Water hyacinth containing heavy metals produces greater amount of biogas. Metals accumulated may be recovered. Treated wastewaters can be used for aquaculture algal culture and for irrigation.

CONCLUSION

Macrophytes based treatment system is a new type of resource - saving and energy saving bio-regenerative technical system in the field of waste treatment. It is a more economic alternative to the common methods like precipitation, sedimentation and ion exchanges. The fundamental difference between conventional and this system is that in the former, wastewater is treated rapidly in highly managed energy-intensive environment whereas in the latter, treatment occurs at a comparatively slow rate in essentially unmanaged natural environment. Development of macrophyte based wastewater treatment system can ultimately be regarded as a first step in the direction of development of more holistic solutions, where waste products will be regarded as a valuable resource which can be utilized for different purposes, instead of being disposed in the cheapest justifiable way.

ACKNOWLEDGEMENTS

The authors are grateful to Dr. V. Vasanthi Devi (Vice-Chancellor) and Dr. A.G. Murugesan acknowledges CSIR, New Delhi.

REFERENCES

Boyd, C.E. (1969). Vascular aquatic plants for mineral nutrient removal from polluted waters. Econ. Bot., **24** : 95-103.

Gain Gupta. (1982). Potential applications of water hyacinth for water air recycling in closed systems. Water Air and Soil Pollution. **17** : 199-205.

Hardy, J.K. and O' Keeffe D.H. 1985. Cadmium uptake by the water hyacinth : effects of root mass, solution volume, complexers and other metal ions. Chemosphere. **14** : 417-426.

Muramoto, S. and Y. Oki. (1983). Removal of some heavy metals from polluted water by water hyacinth. Bull. Environ. contam. Toxicol. **30** : 170-171.

Murugesan, A.G., Vijayalakshmi, G.S., Sukumaran, N., and C. Mariappan. (1994). Utilization of water hyacinth for edible mushroom cultivation. Bioresource Technology, **51** (1) : 97-98.

Murugesan, A.G., Sukumaran, N. and V. Sankar. (1996a). Mushroom cultivation utilizing biological wastes. Proc. Acad. Environ. Biol. **5** (2) : 229-232.

Murugesan, A.G., Priya, P.R., and N. Sukumaran. (1996b). Application of water hyacinth for waste water treatment. Proc. Acad. Environ. Biol. **5** : 72-73.

Murugesan, A.G., Caroline Jeba, W. and N. Sukumaran (1996c). Perspectives of utilizing wastewater for aquaculture after hydrobotanic treatment. Proc. Confaqua '96 : 34.

Murugesan, A.G. and N. Sukumaran (1997). Perspectives of utilizing aquatic macrophytes for wastewater treatment. Indian J. Environ. Protec. **17** (7) : 502-507.

Polprasert, C., Kessomboon, S., and W. Kanjanaprapin. (1992). Pig wastewater treatment in water hyacinth ponds. Wat. Sci. Tech, **26** (9-11) : 2381-2384.

Prasad, B.G.S., Madhavakrishna, W., and Y. Nayudamma. (1983). Utilization of water hyacinth in the treatment and disposal of tannery wastewater. Int. Conf. Water Hyacinth, Hyderabad, p. 56.

Sinha, S.N. and L.P. Sinha. (1969). Studies on use of water hyacinth culture in oxidation ponds treating digested sugar wastes and effluents of septic tanks. Env. Health (UK) **11** : 197-207.

Sridhar, M.K.C. (1988). Uptake of trace elements by water lettuce, *Pistia stratiotes.* Acta hydrochim. hydrobiol. **16** (3) : 293-297.

Srivastav, R.K., Gupta, S.K., Nigam, K.D.P., and P. Vasudevan. (1994). Treatment of chromium and nickel in wastewater by using aquatic plants. Wat. Res., **28** (7) : 1631-1638.

Trivedy, R.K., and V.R. Gudekar. (1985). Water hyacinth for waste-water treatment – A review of the progress. In Current Pollution Researches in India, Environmental Publications, India. 109-145.

Timofeeva, S.S., and D.J. Stom. (1988). Present condition and perspectives of using hydrobotanic treatment for sewage waters. Acta hydrochim. hydrobiol., **16** (3) : 299-312.

Wolverton, B.C., and R.C. Mc Donald. (1975). Water hyacinths and alligator weeds for removal of lead and mercury from polluted waters. NASA Tech. Mem. TM - X - 72723.

20

Anthropogenic Pressures and the Status of the Indian Soft-shell Turtle, *Trionyx gangeticus Cuvier* in Lake Mansar at Jammu

P.L. Duda, Anil K. Verma & D.N. Sahi

INTRODUCTION

The freshwater turtle fauna of India is comprised by 22 species (Das, 1985) belonging to two families viz., the pond turtles, Emydidae (16) and the soft-shell turtles, Trionychidae (6). Both are the wide-spread groups distributed in the Central Region and are directly or indirectly important to man in one way or the other. Freshwater turtles have long served as an important protein resource in tropical and subtropical countries of Asia (Moll, 1985). These are the dominant vertebrates in aquatic habitats and majority of these turtles exhibit herbivore and the nature of diet is advantageous in that there is little competition with man for food and we can therefore, convert a relatively unused resource (Aquatic weed) into a usable form the turtle protein. Another indirect benefit to man concerns, the role played by the turtles especially *Trionyx* spp. in cleansing aquatic habitat as some of the forms are excellent scavengers (Moll, 1985). Soft-shelled turtles are in greatest demand, because of their taste and more fleshy contents when compared to the pond turtles.

AREA OF STUDY

Jammu and Kashmir water bodies are the least investigated for the herptelian fauna and therefore, information about freshwater turtle fauna of the state is relatively meagre (Smith 1977; Duda *et al.* 1977; Gupta 1977; Sahi and Duda, 1985 and Verma, 1992). As a part of studies on Herpeto-fauna of Kashmir Himalayas, supported by D.O.E./M.A.B., New Delhi, a survey of the freshwater turtles of J&K state was undertaken during the years 1988-1992. The river systems of Indus, Jhelum, Chenab and Tawi, the perennial water-bodies like lake Mansar; Surinsar and Dal; the ponds and ditches along the course of the rivers associated reservoirs and nearby village ponds were surveyed for turtles. A description of the survey will be found in the project report (Sahi and Duda, 1990)

J&K State displays a blend of climate and ecology ranging from subtropical plains to temperate hills between 36° 58'N to 76° 20'E longitude. The state is conveniently divisible into three regions viz., Kashmir, Jammu and Ladakh Provinces each with its own climatological characteristics. Turtles were discovered only in the Jammu province of the state. Turtles constituted both the soft-shelled as well as the hard-shelled freshwater forms.

Trionyx gangeticus Cuvier the Indian soft-shelled turtle were recorded both from the lentic as well as lotic water bodies of the Jammu Province. *Trionyx gangeticus* is the most threatened species included under Indian Wild-life (Protection) Act - Schedule I. A sizeable population of the *op. cit.* species inhabits lake Mansar—freshwater eutrophic waterbody 65 km east of Jammu city (32° 42'N - 75° 23'E) at an altitude of 710 mm above sea level in the Shivalik mountains of Jammu Province. This lake is spring fed, heart shaped, 4.3 km in its circumference with the depth of 37 meters surrounded by dense mango tree grooves backed by Pine trees on the slope of the hills. Lake has a thick vegetational cover providing suitable cover for a number of aquatic organisms viz., insects, annelids, molluscs, fishes, amphibians, reptiles and birds. Large population of *Trionyx gangeticus* sympatric with *Lissemys punctata punctata* have been reported earlier from this lake (Sahi, 1978; Sahi and Duda, 1985) but recent surveys by the authors revealed that this lake no longer maintains the large populations of *Trionyx gangeticus.* Present communication reports on the threats to the populations of T. *gangeticus* by multifarious human activities and recommendations for the conservation/management of this turtle species have been made.

ANTHROPOGENIC PRESSURES

The Indian soft-shell turtle population in the lake-Mansar is declining at an alarming rate due to the multifarious human activities or interference in the recent years. The most serious threat to this resource are the habitat alteration or habitat modification. At lake Mansar large tracts of land have been acquired by the authorities to set up various types of Tourist Bunglows and Guest Houses mainly at the bank of the lake which are the breeding sites of the turtle - *Trionyx gangeticus.* Urbanization of the area around the lake and erection of big shopping complexes is another human influence occurring in the population of turtles at the lake which has adversely affected this turtle species. These areas (which were once water catchment areas), where turtle bred during the breeding seasons, provided excellent nesting grounds for the turtle population of the lake. Levelling of these areas has adversely affected the turtle population besides the other flora and fauna of the lake.

Construction of erosion-preventive cemented embankments; fencing of the lake-beaches and beach resortments are the other examples of ecosystem modification of the lake by human interference adversely affecting turtle population at Mansar lake.

Plantations along the lake edge; creation of flowering beds close to the breeding and feeding grounds of *Trionyx gangeticus* and mechanization of ploughing etc. unearths the turtles nests and sometimes the hatchlings and eggs are crushed under the spade work at the lake-edge. Thus, mechanization of agriculture has drastically affected the turtle population in the lake. Moreover, because of over-crowding alongside the lake, turtle feel difficult to come out of water, reach their nesting grounds, to dig the nest and lay the eggs.

Poaching of eggs by villagers fiddling with hatchlings by the school boys seems to be another threat to which the population of lake mansar is being exposed.

Raising of the lake resorts, non-mechanical and mechanized boats, vehicular traffic alongside of lake margins; artificial lighting in the vicinity of major mass nesting grounds are undoubtedly posing most serious threats to the turtle population at lake Mansar leading to the situation that, these turtles seems to be on their way out which has struck the authors. Further, proliferation of non-mechanised and mechanised fishing boats have increased the operational range and efficiency of the fishermen making the entire area of the water body once remote, easily accessible. Activity of large number of fishing boats and tourists boats and use of nets often create artificial barriers in the waterbody preventing the turtles to emerge on the nesting ground for nesting.

Besides all the above mentioned human interferences, other threats like exploitation of the soft-shell trutle *Trionyx gangeticus* by man for eggs, meat and other by-products have caused depletion of turtle population in lake Mansar. Apparently the exploitation of the turtle species, *Trionyx gangeticus* over the years, habitat alterations at the lake and heavy loss in the eggs of these turtles seems to be contributing to the small population-size of this turtle species in lake Mansar.

PRESENT STATUS AND RECOMMENDATIONS

Obviously an attention both in magnitude and urgency is deserved by these important faunal elements which have been included in the Schedule-I of the Indian Wild Life (Protection) Act, 1972. Their conservation and survival is very important because, they constitute an important component of the ecosystem. Following are the recommendations for conserving the population of *T. gangeticus* in lake Mansar :

1. Promulgation of law enforcing regulations on fishing/ Predation on the turtles for meat and eggs.
2. An yearwise assessment of the population-size and locating the turtle's nest and their protection from other predators.
3. Establishing the area in the vicinity of the lake as the sanctuary.
4. Raising of the hatcheries near the lake site and switching over to the captive breeding programmes.
5. Prohibition of the collection of turtles especially during their nesting season.
6. Incentive for protecting turtle nests should be given to the local men to create interest among them for the turtles.
7. Educating the common man about the significance of these creatures in the ecosystem.

It is strongly recommended that no more Tourists Resorts and Recreational Centres should be raised near the lake by the Department of Tourism. The Wildlife Department should have a strong vigil round the clock and the department should depute personnel to disseminate educational information to the public concerning game-laws and conservation programmes.

"Thus, an awareness needs to be created to make sure we help our turtle population to survive and thrive."

ACKNOWLEDGEMENTS

Authors are grateful to the Department of Environment and Forests (M & B), Govt. of India, New Delhi for the financial assistance provided under the Project, "Herpeto-Fauna of Kashmir Himalayas etc...."

REFERENCES

Das, I. 1985. Indian Turtles : A Field Guide. World Wildlife Fund India, Eastern Region : 1-119 Ms. Bonani Kakkar (Publ.).

Duda, P.L., Sahi, D.N. and Dutta, S.P.S. 1977. On the occurrence of some tortoises of Jammu. *Geobios.*

Gupta, V.K. 1977. Studies on the female reproductive cycle of some chelones form Jammu. Ph. D. Thesis, University of Jammu, Jammu.

Moll, E.O. 1985. Freshwater turtles in India, their status, conservation and management. *Hamdaryad,* **9** (3) : 9-17.

Sahi, D.N. 1978. A contribution to the herptiles of Jammu & Kashmir State. Ph.D. thesis, University of Jammu, Jammu.

Sahi, D.N. and Duda, P.L. 1985. Checklist and keys to the Herptiles of Jammu & Kashmir state. *Chic. Herp. Soc.* **20** (3-4) : 86-97.

Sahi, D.N. and Duda, P.L. 1990. Herpeto-fauna of Kashmir Himalayas : Exploration, Ecology and Conservation. Project Report. D.O.E. MAB, New Delhi.

Smith, M.A. 1931. "The Fauna of British India, Including Ceylon and Burma". Reptilia and Amphibia 1. Loricata, Testudines Taylor and Francis (Eds.), London.

Verma, A.K. 1992. Studies on the male reproductive cycles of some emydid turtles of Jammu. Ph.D. thesis, University of Jammu, Jammu.

Index